SYSTEMATICS AND BIOLOGY OF THE GENUS *MACROCNEME* HÜBNER (LEPIDOPTERA: CTENUCHIDAE)

Systematics and Biology
of the Genus *Macrocneme* Hübner
(Lepidoptera: Ctenuchidae)

Robert E. Dietz IV

UNIVERSITY OF CALIFORNIA PRESS
Berkeley • Los Angeles • London

Volume 113
Issue Date: April 1994

UNIVERSITY OF CALIFORNIA PRESS
BERKELEY AND LOS ANGELES, CALIFORNIA

UNIVERSITY OF CALIFORNIA PRESS, LTD.
LONDON, ENGLAND

Library of Congress Cataloging-in-Publication Data

Dietz, Robert E. (Robert Edwin), 1941–
    Systematics and biology of the genus Macrocneme Hübner
(Lepidoptera: Ctenuchidae) / Robert E. Dietz IV.
        p.   cm. — (University of California publications in
entomology; v. 113
    Includes bibliographical references (p.   ).
    ISBN 0-520-09780-7 (pbk. : alk. paper)
    1. Macrocneme.   I. Title.   II. Series.
QL561.C8D54   1994
595.78'1—dc20   93-49869
                                                       CIP

Contents

Acknowledgments and Abbreviations

Much of the material on which this study is based was borrowed from or examined at museums in the United States, Latin America, and Europe. I thank the curators who assisted me in arranging loans, answering questions, and/or providing facilities for the study of types and specimens in their charge.

Listed below are the institutions and individuals who cooperated in this study, along with the abbreviation used for each throughout the text:

AMNH American Museum of Natural History, New York. (F.H. Rindge)
BMNH British Museum (Natural History), London. (A. Watson; in Glasgow, R.A. Crowson)
CAS California Academy of Sciences, San Francisco. (P.H. Arnaud)
CM Carnegie Museum, Pittsburgh, PA. (H.K. Clench)
CNC Canadian National Collection, Ottawa. (E.G. Munroe)
CU Cornell University, Ithaca, New York. (L.L. Pechuman)
HM Zoologisches Museum, Universität Hamburg, Germany. (H. Weidner)
IML Fundación Miguel Lillo, Tucumán, Argentina. (A. Willink)
LACM Los Angeles County Museum of Natural History, Los Angeles, California. (J.P. Donahue)
MCZ Museum of Comparative Zoology, Harvard University, Cambridge, Massachusetts. (J.M. Burns)
MNHU Museum für Naturkunde der Humboldt-Universität, Berlin. (H.J. Hannemann)
MSU Michigan State University, Lansing. (R.L. Fischer)
NMB Naturhistorisches Museum, Bern, Switzerland. (H.D. Volkart)
NMG Muséum d'Histoire Naturelle, Geneva. (C. Besuchet)
NRS Naturhistoriska Riksmuseet, Stockholm. (B. Gustafsson)
OX Oxford University Museum, Oxford, England. (E. Taylor)
PM Muséum National d'Histoire Naturelle, Paris. (P. Viette)
RML Rijksmuseum van Natuurlijke Historie, Leiden, the Netherlands. (R. de Jong)
SMM Zoologische Sammlung des Bayerischen Staates, Munich. (W. Förster)
UCB University of California, Berkeley. (J.A. Powell)
UCD University of California, Davis. (R.O. Schuster)

UCV	Universidad Central de Venezuela, Maracay. (F. Fernández-Yépez)
UFP	Universidade Federal do Paraná, Curitiba, Brazil. (O.H.H. Mielke)
ULP	Universidad Nacional de La Plata, Argentina. (L. De Santis)
UPP	University of Pennsylvania, Philadelphia. (D.H. Janzen)
USNM	United States National Museum [National Museum of Natural History], Smithsonian Institution, Washington, D.C. (E.L. Todd)
USP	Museu de Zoologia, Universidade de São Paulo, Brazil. (N. Bernardi)
VM	Naturhistorisches Museum, Vienna. (F. Kasy)
ZIH	Martin Luther Universität, Halle, Germany. (Hüsing)

Without the assistance and encouragement of numerous friends and colleagues during the long course of this study, this project might never have been completed. I owe special thanks to Jerry Powell, major advisor, tutor, and friend, for his patience and unstinting assistance in seeing this study through to completion. His guidance was responsible for numerous improvements in the content and presentation of this paper.

Robert Ornduff and Evert Schlinger, as additional members of the dissertation committee, reviewed the manuscript. Their suggestions and criticisms are gratefully acknowledged.

W. Donald Duckworth and [the late] Harold Moore gave me valuable experience in the tropics by including me as an assistant on their collecting expeditions. This project was begun under their tutelage, and their unfailing support and interest in my progress will always be remembered and appreciated.

Allan Watson was especially helpful in returning types to me for further study and in answering those inevitable questions that arise in a protracted project where early details have a way of fading from memory. Frederick Rindge kindly reexamined the type of *spinivalva* Fleming for me. [The late] Francisco Fernández Yépez helped arrange for my visit to Maracay in 1974-75. His innumerable kindnesses and assistance during that year are warmly remembered.

Michael Adams, George Bernard, and Michael Murtaugh suffered the vagaries and frustrations of collecting in the Tropics with me. Their forbearance, humor, and good company in the face of plagues of swarming insects made all bearable.

Among those who contributed specimens to this study from their own collecting efforts or searched for material on my behalf are: Yves Barbotín (Cayenne, French Guiana); André Blanchard (Houston, Texas); John Chemsak (Berkeley); Rosser Garrison (San Juan, Puerto Rico); George Godfrey (Champaign, Illinois); John Hafernik (San Francisco); Daniel Janzen (Philadelphia); Gerardo Lamas (Lima, Peru); [the late] Harold Moore (Ithaca, New York); Tom Pliske (Coral Gables, Florida); Jerry Powell (Berkeley); Don Viers (Pullman, Washington); Philip Ward (Davis, California).

Dale Habeck provided information about rearing ctenuchids, and Norman Leppla helped to reassess the artificial diet. Their suggestions were useful in improving the rearing results.

Special thanks are due the following for their assistance during various phases of the study: friends Phillip Guins and Alden Kamikawa helped with labeling; Richard Clarke collated locality data, helped prepare maps, and proofread various versions of the manuscript; Hinrich Seeba and Martha Bell Dietz translated articles from German; John DeBenedictis took the SEM photographs; Dean and Chalmers Luckhart took the photographs of the adults; Karen Bailey helped with the typing; Mary Anne Clarke assisted me in converting hand-drawn graphs to computer-generated ones.

To my mother and late father, Ann and Robert Dietz, I owe special thanks for their generous support and encouragement over the course of this study. Their unflagging faith and confidence in me were invaluable morale boosters.

This study was aided in part by the following grants and awards: National Science Foundation Grant 6813X (to Jerry A. Powell, 1969-70); National Institutes of Health Training Grant #AI-218, UCB; Chancellor's Patent Fund Award, UCB; Predoctoral Fellowship Award, Office of Academic Studies, Smithsonian Institution; Organization of American States Research Fellowship #PRA-41437. Final publication costs were defrayed, in part, by a generous gift from Ann Espe Dietz.

Note: Present address of author: Dr. Robert E. Dietz IV, 6939 Edith Boulevard, NE, Albuquerque, New Mexico 87113.

Abstract

Thirty species considered monophyletic and congeneric with *Macrocneme maja* (F.) are characterized, illustrated, and discussed. Fourteen new species are described from South America: *ancaverdia* (Peru); *bestia* (Brazil); *bodoquero* (Colombia); *durcata* (Venezuela); *habroceladon* (Bolivia); *imbellis* (Peru); *megacybe* (Brazil); *melanopeza* (Peru); *mormo* (Brazil); *oponiensis* (Colombia); *orichalcea* (Brazil); *pelotas* (Brazil); *tarsispecca* (Bolivia); *zongonata* (Brazil). Three names are removed from synonymy, one subspecific name is reinstated as a valid species, and nine names are newly placed in synonymy. Three species erroneously associated with *Macrocneme* are transferred to other genera. Biological information is summarized including a new hostplant record for larvae collected in the wild and four new larval descriptions based on rearings with artificial diets. Patterns of geographical distribution are discussed and illustrated. Parallel geographic variation is suggested for various species groups involved in Müllerian mimicry complexes.

INTRODUCTION

Macrocneme is a Neotropical genus of 30 day-flying moth species that are readily recognizable by their striking mimetic resemblance to pompilid and ichneumonid wasps. This study concludes a revision originally begun during a review of the genus *Horama* where it was discovered that species in *Horama* and *Macrocneme* had been mixed, even though the genera were in different subfamilies (Dietz and Duckworth, 1976). A revision was undertaken to redefine the generic parameters and to describe the included species. Species had been classified more by superficial resemblance than by phyletic kinship. Generic placement was often impossible due to the effects of convergence, mimicry, and variation (polymorphic, polytypic, or geographic).

At the conclusion of the *Horama* review, 29 species-names remained in *Macrocneme* sensu stricto. I have redefined the genus on the basis of the morphology of the genitalia, especially the male structures. Accurate identification of species by sight is difficult, owing to the similar appearance of the adults and to sympatry among numerous species. Reliable identifications are possible only by examining the genitalia. Forbes (1939, in part) reviewed *Macrocneme* and made the first attempt to compare features of the genitalia. His observations and nomenclatorial changes were made without an examination of the types; the inaccuracies that resulted have been addressed in this study. I also summarize all available biological information, including the results of a year's fieldwork, and discuss the geographical distributions.

MATERIALS AND METHODS

Approximately 4500 specimens of adults representing 30 species of *Macrocneme* were examined in this study. Material was seen from all the major collections in North America, Europe, and South America (see Acknowledgments). Not all species were well represented, due partially to the difficulty in associating the sexes and partially to meager collections of the lesser-known species. I obtained eggs of 8 species, and laboratory-reared on an artificial diet (with varying degrees of success) larvae representing 5 species (see Biology and Appendix). A general diagnosis for a larva of *Macrocneme* and a description of the chaetotaxy (including a setal map) resulted from this material. Full-grown larvae and pupae are newly described for 3 species.

Differences in the morphology of the male genitalia are the principal means for distinguishing among species. Thus males were used to describe 3 new species with unknown females, but no new species are based solely on females.

Genitalia were dissected from 221 specimens (134 males, 87 females), using the usual KOH procedure for preparing Lepidoptera genitalia. The preparations were stored in glass vials containing glycerin and stoppered with polyethylene plugs. Labels with matching numbers were pinned to the vials and to the corresponding specimen from which the genitalia had been removed. It is perhaps not advisable to permanently store the preparations in glycerin, since with the passage of time the membranes become fragile and easily torn.

The glycerin procedure allowed me to view the asymmetry of the various parts without the additional distortion caused by a slide mount. The number of preparations varied, depending on my familiarity with the species and the availability of material, and are noted in the descriptions. Diagnostic features of the male genitalia were usually visible without the need for further dissection. Only the aedeagus required removal for accurate viewing. For each species I have illustrated the four diagnostic features used in identifying a specimen: left valva, dorsum of the uncus, juxta, and aedeagus. The extent to which the juxta extended beyond the base of the ventral processes on the valvae was often diagnostic. In the illustration of the juxta I have indicated with paired dashes [– –] the relative position of the base of the ventral processes.

It always required a dissection to view the intersegmental modifications and the 7th sternite in females. To view the internal organs, the inner fold of the 7th sternite

2

had to be separated from the posterior segments along a firmly attached ostial membrane. Care had to be taken not to tear between the sclerite of the lamella postvaginalis (intersegmental) and the two caudal segments (8, 9 + 10). If a separation was made at this point, the internal organs from the ostium to the corpus bursae remained attached to the pelt.

I have illustrated and labeled only one example of the internal features of the female genitalia (Fig. 134) to aid in the text descriptions. The general structure is the same for all species. The shape of the seventh sternite and accompanying intersegmental modifications were found to be reliable species indicators. These structures have been illustrated for all species where I was reasonably confident the females were properly associated (Figs. 146-188).

Wing measurements were made from five individuals of each sex where material permitted. Wing slides were prepared for four representative species, but the venation was found to be too variable to draw any useful conclusions regarding interspecific differences.

The tymbal organ was photographed for four species under a scanning electron microscope. Differences in microtymbal hair scales are shown in Figs. 195-200. The procedure required for photographing these organs involved essentially destroying the specimen so I looked only at representative species where material was abundant.

Drawings were made from individual preparations with the aid of a camera lucida, as noted in the descriptions, and are not composite. Size was not considered in my assessment of the genitalic structures. Thus larger specimens are drawn to smaller scale and figures of different enlargement appear together. As an example, the aedeagus and sclerite of the lamella postvaginalis are 2X the size of other figures appearing together on the same page. Institutional deposition of the illustrated specimens is noted after each preparation number in the text.

Species descriptions are each based on a representative specimen from the available material and are not composite. Similarly, genitalia descriptions are taken from the illustrated preparations and are not composite. Variation is summarized for both sexes following the description of the female.

Data for the specimens examined is given in the order of: country, state or province, date, elevation, and collection. I have added the state, province, or department if it is not given on the label. Dates are given by months only and summarized in the flight-period information. All localities are shown in species maps (Maps 4-27). For the sake of economy I have summarized the specimens examined and have included collector names only for type material and for questionable localities. Corrected spellings and other changes are included within brackets, [ ]. The following references were particularly useful in clarifying many of the old and obscure localities: F.M. Brown (1941); Selander and Vaurie (1962); Fairchild and Handley (1966); Lamas (1976); and K.S. Brown (1979). K.S. Brown's appendix (1979) is an excellent gazetteer for old Lepidoptera collection localities in Central and South America.

I was able to locate the original type material for all names used in this study except *chionopus* Draudt, *chrysitis* Guérin, and *leucostigma* Perty. Dr. Schroeder at the Senckenberg Natur-Museum in Frankfurt suggested (pers. comm.) that the Draudt type was probably lost during World War II when the collection was destroyed.

An exclamation point (!) before the name indicates I have examined the type specimen. A label has been attached to all specimens I have identified.

By far the largest collection of adults is located in London (BMNH). Other excellent representations are to be found in Munich (SMM) and Washington, D.C. (USNM).

No attempt has been made to include all species citations in the literature references. The taxonomy of the genus has been in such a confused state for so long that many of the names used in the literature are unreliable. I have attempted to cite all generic changes and any references to subsequent usage of a name when it was clear the authors were referring to the same species. Since the most reliable identifications were based on features of the male genitalia, I have illustrated appropriate characters and included a key. Identification of properly associated females may sometimes be problematical, since data for all species is incomplete. I have illustrated characters in the female genitalia which were useful in separating species. A comparison of these figures with a specimen at hand should identify a species as I envisioned it.

NOMENCLATURE AND TAXONOMIC HISTORY

Hübner established *Macrocneme* in 1818 with two included species, *maja* and *lades*. No author was given for *maja*, and *lades* was cited from a Fabrician work (1781) without reference to Cramer. In the *Verzeichniss* Hübner corrected these oversights (1819), with *maja* clearly attributed to Fabricius and *lades* to Cramer. There is no question which species Hübner was transferring to *Macrocneme*, and I consider his unattributed *maja* to be a *lapsus* rather than a primary homonym.

The generic synonymy for *Macrocneme* is not complicated. Only two other genera, *Copaena* Herrich-Shäffer and *Poliopastea* Hampson have been associated with the genus. At present no genera stand in synonymy. Kirby (1892) considered *Copaena* a synonym of *Macrocneme* on the basis that *maja* F. was the type of *Macrocneme*. Many of the plates for Herrich-Shäffer's work (1850-[1869]) were published after the text. *Copaena* appeared first in combination with *scapularis* Herrich-Shäffer on the dust wrappers for the plates (1855) and then in combination with *maja* in the text (1856). The earlier combination makes *scapularis* the type of *Copaena* by monotypy. Most recently, Field (1975) has considered *Copaena* to be a junior synonym of *Antichloris* Hübner, which is not closely related to *Macrocneme* according to the male genitalia.

Fleming placed *Poliopastea* in synonymy with *Macrocneme* in 1957, dismissing as variable the palpal and venational characters which Hampson had used to separate the two genera. He formally transferred only *P. plumbea* Hampson to *Macrocneme* — but since it was the type, all included species were presumably also to be considered in *Macrocneme*. There were five: *coelebs* Bryk, *obscura* Wallengren, *pava* Dognin, *vidua* Bryk, and *viridis* Druce. Besides *plumbea*, only *coelebs* and *obscura* were formally returned to *Poliopastea* when the genus was reestablished (Dietz and Duckworth, 1976).

M. pava, *vidua*, and *viridis* were anomalous, and were left in *Macrocneme* pending further study. With the completion of the present review it is apparent that none should be considered in *Macrocneme*, and they have been formally removed. (see Species Transferred from *Macrocneme*).

At the time of the last ctenuchid catalog (Zerny, 1912), 42 species names were associated with *Macrocneme*. In the interval between 1912 and 1957 (last new species described), 16 names were added. Fleming's synonymy of *Poliopastea* (6 species) brought the total to 64. Aside from the three anomalous species mentioned above, the placement of these names fell naturally into three genera. *Poliopastea* Hampson was reestablished with 33 names transferred from *Macrocneme*, and *viridifusa* Schaus and *nigricornis* Schrottky were associated with *Horama* (Dietz and Duckworth, 1976), while 29 names remained in *Macrocneme (sensu stricto)*. The disposition of these last 29 names within the body of the text is as follows:

11	valid species:	*adonis* Druce, *aurifera* Hampson, *chrysitis* Guérin, *coerulescens* Dognin, *cupreipennis* Walker, *cyanea* Butler, *immanis* Hampson, *lades* Cramer, *leucostigma* Perty, *thyra* Möschler, *thyridia* Hampson
9	new synonyms:	*aurata* Walker, *boliviana* Draudt, *chionopus* Draudt, *chiriquicola* Strand, *cinyras* Schaus, *deceptans* Draudt, *intacta* Draudt, *spinivalva* Fleming, *yepezi* Förster
4	continuing synonyms:	*affinis* Klages, *albiventer* Dognin, *euphrasia* Schaus, *guyanensis* Dognin
3	names removed from synonymy:	*ferrea* Butler, *iole* Druce, *semiviridis* Druce
1	subspecific name removed from synonymy and reinstated as valid species:	*cabimensis* Dyar
1	type species:	*maja* F.

In addition, *chlorata* Dognin was found to have been misplaced in *Calotonos*. It is treated here as a synonym of *thyra*. For species previously undescribed, 14 new names have been proposed, bringing the total number of species considered valid to 30.

GENUS *MACROCNEME* HÜBNER

Macrocneme Hübner, 1818:15, pl. [12], figs. 65 and 66
Euchromia (Macrocneme), Walker, 1854:248
Type species: *Zygaena maja* Fabricius, 1787, by subsequent designation, Kirby, 1892.

DIAGNOSIS AND SYSTEMATIC PLACEMENT

This genus is a remarkably uniform assemblage of blackish moths with blue to green iridescent wings that presumably mimic pompilid wasps. The color and pattern of iridescence, while diagnostic for some species, is too variable for reliable species recognition. The genitalia should always be examined for species determination.

The phyletic affinities are uncertain. Resemblance in overall similarity of the genitalia and in the external appearance suggest a monophyletic origin. Species pairs and assemblages are apparent (see sections below on Parallel Geographic Variation and Geographic Distribution), but are probably due to polymorphism, polytypy, or mimicry rather than phyletic kinship.

The phenotype is essentially the same for every species. The ground color is brownish black, the hind legs are long and plumose, the wings are variously patterned with blue and/or green iridescence. All species possess white spots on the tips of the antennae, on the frons, at the base of the labial palpi, at the base of the wings (above and below), on the abdominal venter and pleura, and on the first abdominal tergite.

The male genitalia have two processes on the valvae separated by a mesal sclerite. One is the clasper arm (dorsad) and the other is a soft, setose rod (ventrad). Most species have two aedeagal spines, the exceptions being *thyridia* with three; *adonis, cabimensis, durcata, chrysitis,* and *iole* with one; and *semiviridis* with a spatulate process rather than a distinct spine. The lateral margins of the uncus are flanged, although narrowly in *pelotas* and *thyridia*.The uncus is often asymmetrical, as are the tips of the clasper arms. The juxta is spined at the apex, except in those species where the distal margin extends only to the base of the ventral processes on the valvae, e.g., in *aurifera, durcata, immanis,* and *megacybe*. Only *thyra* has a short juxta with a spiny patch at the left margin.

In the female genitalia, the sterigma includes a modified seventh sternite in which the intersegmental areas between segments VI-VII and VII-VIII are variously sclerotized. The anterior apophyses are fused laterally into bulla-like structures. The lamella postvaginalis is an intersegmental sclerite between segments VII and VIII. There is always an accessory bursa, and the corpus bursae contains two scallop-shaped, spiny signa.

A striking similarity in habitus with species in *Calonotos (C. verdivittata), Saurita (S. bipuncta),* and *Pseudosphenoptera (P. triangulifera)* suggests that the closest ally is a genus that remains to be defined. Aside from genitalic differences, these species differ principally in having hyaline spots on the wings and lacking enlarged hind legs. Even though *Poliopastea* Hampson has long been associated with *Macrocneme,* its different abdominal morphology, including an unusual "ventral valve," suggests that a close evolutionary relationship is not likely (see Dietz and Duckworth, 1976).

DESCRIPTION

Last Instar Larva (see Figs. 189-192)

Length 20-35 ± mm; head yellow, black, or with prominent dark mesal band; all segments with dark plumose setae in tufts; cuticle ground-color grey to black; segment margins often bordered with yellow or spotted; verrucae bases well sclerotized, often iridescent blue or green; dense black verricules on segments A1 and A7; long white setae from T3 and A8; extra-long, black setal tufts projecting forward from T2. See *thyra, coerulescens,* or *thyridia* for additional details.

Larval Chaetotaxy (see Fig. 4)

Cervical shield with secondary setae to either side of XD1 and XD2 along margin; D1 and D2 of shield without secondary setae; D1, D2, and SD2 combined as single verruca (D + SD) on T2 and T3; dorsal verrucae D1 small and contiguous on A1 but separate on A2-A8; verruca D2 prominent on T3-A8; D and SD joined on A9 to form large verruca; SD1 and SD2 from single pinaculum on T1; SD1 of T2 and T3 associated with small pigmented pit ventral to basal socket; SD1 and SD2 fused on A1-A9 forming verruca SD; L1-L3 fused on T1-T3 and A9 forming verruca L; L1-L3 forming separate verrucae on A1-A8; verruca L1 usually small but enlarged and modified as black tufted verricules on A1 and A7; verruca L2 often slightly larger than verruca L1 and modified as verricule on A1; verruca L3 prominent and larger than either L1 or L2; verruca SV small but distinct on A1, A2, A7-9, but indistinguishable from setae on prolegs plates of A3-A6; verruca V tiny, basomesad of coxae, united on T1, separate on T2 and T3, slightly larger and distinct on A1, A2, A7-9; abdominal crochets 15-19; anal crochets 13-16.

Adult

Head: Brownish black to black. Eye diameter of males averages 0.2 mm larger than females. Antennae in both sexes bipectinate becoming dentate distally, scaled dorsally, tips white, pectinations longer in males than in females. Frons smooth with white spot below each antenna. Pilifers well-developed. Ocelli present. Occiput metallic scaled. Maxillary palpi minute, 1-segmented. Labial palpi 3-segmented, upturned or porrect, occasionally reaching antennal bases; segment I rough-scaled with basolateral white spot; segment II smooth or with slightly raised scales on inner surface, occasional white irrorations on outer surface; segment III short, sometimes obscured by scales.

Thorax: Brownish black to black, with iridescent markings on disc, pectus, tegulae, and patagia; thin, non-metallic, hairy scales overlie and obscure iridescence on disc and pectus. Tymbal organ present on anterior margin of metepisternum. White spots always on subdorsal and sublateral edges of patagia, on proximal, mesal, and lateral margins of forecoxae, on meso- and metapleura, and trochanters; white spots frequently on propleuron, mid- and hindcoxae, and underside of tegulae. Legs, especially forecoxae and all tibiae, streaked with iridescent scales; hind legs appearing enlarged due to elongate scales in double rows on tibiae and a single row on tarsi; tips of hind tarsi either black or white. Forewing length, 14-24 mm. Wing ground-color usually brownish black, sometimes more strongly black or brown depending on the species, population, or age of specimen; iridescent scales prominent, but pattern and color variable; forewing with two basal white spots above and one below; retinaculum often white; hindwing with ill-defined white patch at humeral angle above, small distinct spot below; in hindwing Cu_1 and Cu_2 united but forked at termen, M_2 rudimentary with only trace of vein from termen.

Abdomen: dark green to black with dorsum usually suffused with metallic green or blue; iridescence dull or shiny, either entirely covering dorsum or as three thin, longitudinal stripes; iridescence of venter restricted to lateral margins; basal tergite brownish black to black, with paired subdorsal and sublateral white spots and an occasional small metallic dot at center; prespiracular, counter tympanal hoods forming lateral bullae on basal tergite; basal pleurite white-spotted, remaining pleurites with spots variable; venter usually with mesal series of white spots, often obsolescent caudally, or entirely white; bilobed intersegmental invagination (gland?) between sternites VII and VIII in males.

Male Genitalia (see Fig. 5): prominent, heavily sclerotized; valvae simple with 2 dorsocaudal processes, the uppermost the strongly sclerotized "clasper" arm, variously shaped, usually asymmetrical, the lowermost weakly sclerotized, rod-like, occasionally recurved, with long setae sometimes interspersed with long-shafted scales; mesal sclerite between clasper and lower process often with tiny protuberance; juxta and dorsum of uncus usually asymmetrical, with diagnostic shape for each species; saccus absent; dorsum of aedeagus sclerotized beyond attachment to diaphragma and flat sclerotized, with 1-3 spines laterally; vesica of 3 membranous bursae with some sclerotization at origin of apical bursa.

Female Genitalia (see Figs. 134 and 135): sterigma formed by modifications in sternites VII and VIII and associated intersegmental membranes; shape of sternite VII often diagnostic, generally scutiform or inversely triangulate, with apical margin often skewing to left of midline; pleurites encircle from sides; intersegmental cuticula between VI and VII often modified, either as a broad, sclerotized depression, as well-defined pockets (often in pairs), or as small, irregular sclerites; ostium encircled with lamella antevaginalis, the sclerotized inner and outer folds of sternite VII and lamella postvaginalis an irregularly ovate sclerite between segments VII and VIII, often decurved at ostial opening; sternite VIII strongly sclerotized with a prominent protuberance and usually 2 thickened plicae at center, lateral margins terminate in bulbous pockets that serve as the anterior apophyses; ductus bursae short, broad, usually nonsclerotized; small blind pouch or thickened plicae often evaginate from dorsal wall of ductus bursae; accessory bursa at right, long-stalked, with wide, geniculate base; ductus seminalis usually arising dorsocephalad near inception of accessory bursa; bulla seminalis present; corpus bursae membranous, usually in concentric plicae; signa always of 2 opposing, scallop-shaped patches with recumbent spines.

VARIATION

All species show individual and geographic variation in the color and pattern of wing iridescence. Within a population the color can vary from blue to green or be a mixture of these colors. In the forewing, the metallic scales may suffuse over the whole wing or be confined to the basal half, with the iridescence entire or interrupted by a variable black streak from the basal angle. In the hindwing the discal area may have iridescent scales present or absent depending on the population.

Clinal variation is evident in *Macrocneme durcata* and *cabimensis* and will probably show up in other species as they become better known.

Only one species is sexually dimorphic in color and pattern of wing iridescence. In females of *M. adonis* the iridescence is bright blue and reaches the apex, whereas in males it is green and seldom extends beyond the end of the cell.

There is some variation in the shape and size of various parts of the male genitalia, especially the juxta, clasper arms of the valvae, and the aedeagal spines. The shape of the 7th sternite, especially the caudal margin, varies in females, but not sufficiently to prevent recognition. The number of spermatophores found at the base of the accessory bursa varies between 1 and 3.

Males are invariably whiter than females on the underside of the thorax and abdomen. Whenever the pectus, proximal leg segments, and abdominal venter are strongly white in males, the corresponding markings in females are usually reduced to small white spots or absent.

BIOLOGY

While the adults of *Macrocneme* are readily recognized in the field and seldom ignored by collectors, little of their natural history is recorded. Most published observations comment primarily on the mimetic resemblance of the genus to pompilid wasps in genera such as *Pepsis, Pompilius,* or *Salius*[= *Cryptocheilus*] (Seitz, 1890; Schrottky, 1909; Kaye, 1913; Moss, 1947). More recent studies have treated adult habits, palatability, and warning displays in selected species (Beebe and Kenedy, 1957; Blest, 1964).

Only two accounts are available concerning the early stages. The earliest, by Mabilde (1896), describes the larva of *M. iole* and a later one by Jorgensen (1913), describes that for *lades*. Unfortunately, *iole* does not occur in Brazil, so the identity of the Mabilde larvae is uncertain. Both descriptions appear to treat the same species, and I consider them here as one under *leucostigma* Perty.

Host Associations

Larval host-plant associations are scarce. Mabilde (1896) records *M. iole* on two Brazilian plants identified by Costa Lima (1968) as *Lantana sellowiana* [Verbenaceae] and *Mikania amara* [Compositae].

Recently, Daniel Janzen (unpub.) has reared *Macrocneme lades* from larvae collected in the wild on vines of *Mesechites trifida* [Apocynaceae] in Santa Rosa National Park, Guanacaste, Costa Rica (voucher tag 89-SRNP-340 [UPP]). This is the first record of a *Macrocneme* reared on this family, well-known for its poisonous secondary substances. (see section below on Defense for further discussion). All other host associations refer to adults which have been collected or observed visiting the plants listed in Table 1.

A wide variety of ctenuchids is attracted to these plants (Seitz, 1916). Most likely in *Macrocneme* they serve only as nectar sources for the adults. The visits are infrequent, which suggests the associations are facultative. In other ctenuchids like *Poliopastea*, visits by males to these plants are presumably required as a prelude to the production of pheromones used in mating (see section on Plant Attraction). There is no evidence that *Macrocneme* adults have similar requirements. I suspect these plant genera have little in common with the larval food-plants for *Macrocneme*.

11

Table 1. Inflorescences Visited by Adults of *Macrocneme.*

Family	Genus [a]	*Macrocneme* Visitor species	Source
Boraginaceae	*Cordia*	*lades*	Opler (unpub.)
	Heliotropium	*adonis, lades*	Pliske (1975)
Compositae	*Baccharis*	? *leucostigma*	Jorgensen (1913)
	Baltimora	*lades*	Opler (unpub.)
	Eupatorium	? *leucostigma* [b]	Jorgensen (1913)
	Melanthera	*lades*	Opler (unpub.)
	Moquinia	? *leucostigma* [b]	Jorgensen (1913)
	Senecio	? *leucostigma* [b]	Jorgensen (1913)
	Veronia	? *leucostigma* [b]	Jorgensen (1913)
Leguminosae	*Acacia*	? *leucostigma* [b]	Jorgensen (1913)
Rubiaceae	*Genipa*	*lades*	Opler (unpub.)
Zingiberaceae	? [c]	"sp. nov."	Beebe & Fleming (1951)

[a]. For species names, see Adult Host Records under *M. lades* and *M. leucostigma.*
[b]. Cited as *lades*, a probable misidentification.
[c]. Cited as "ginger blossoms."

Rearing

Because information about the early stages of *Macrocneme* was scarce, I attempted to rear various species on an artificial diet. I included arctiids as well as other genera of ctenuchids in order to compare differences in acceptability and rates of development between these closely related groups.

Procedure: I began by collecting females at light and placing them in plastic bags. The bags were stored on shelves in an air-conditioned laboratory(24° C) at the Department of Entomology, Universidad Central de Venezuela, in Maracay. Also included were leaf samples from various plant genera (see *adonis* Biology for names) and a small, cotton-stoppered vial with a weak solution of honey water. While the females showed no preference for the leaves, they often probed the moist cotton with their tongues. At oviposition the eggs were cut from the bag and placed on a diet contained in plastic "creamer" cups. Plant samples were initially included with the diet to test for possible preference on a natural host. This procedure was discontinued when the larvae did not move to the leaves, and no feeding was observed. Upon hatching, the larvae were kept together until the first moult and then divided into lots of 3 to 5 individuals depending on their rate of development. Moults were determined by the appearance of head capsules. These were removed with a moist brush and stored in sequence on stripes of sticky tape. The diameters were subsequently measured to determine the number of instars. The larvae spun flimsy cocoons on the lid or side of the cups before pupating. The cocoon and enclosed pupa were placed in screened cages to await emergence of the adult.

A broad-spectrum meridic diet was adapted from several proposed by Shorey and Hale (1965) and Schroeder (1970). (See Appendix for contents and method of preparation.) Lima beans were experimentally substituted for pinto beans without any noticeable difference in acceptability by the larvae. Processed (i.e., toasted) wheat germ was supplemented with linseed oil to provide linolinic acid, an essential fatty acid in some Lepidoptera diets for proper emergence from the pupa (Vanderzant et al., 1957). Formaldehyde was added as a preservative and the potassium sorbate, methyl paraben, and tetracycline were included as mould inhibitors. Drying and the growth of surface bacteria necessitated replacing the diet at 5/6-day intervals. Increasing the water content and mould inhibitors proportionally by a third made it possible to maintain the larvae on the same diet for 12 days. The surface microorganisms were virtually eliminated by immersing the presoaked beans in boiling water and then removing them from the heat once the water reboiled. This procedure minimized the leaching of minerals and the destruction of the labile fats and proteins.

Results: 11 ctenuchid and 8 arctiid species were reared to various stages of development. The ctenuchids which produced adults were: *Gymnelia flavitarsis* Walker, *Hypocharis clusia* Druce, *Macrocneme thyra* Möschler, *M. coerulescens* Dognin, *Poliopastea nigritarsia* Hampson, and a *Saurita* sp. Final instars were seen for: *Macrocneme thyridia* Hampson and *Poliopastea chrysotarsia* Hampson. Others which initially accepted the diet but succumbed by the 3rd instar were: *Macrocneme durcata* n. sp., *Histiaea cepheus* Cramer, and *Histiaea* sp.

The arctiids which produced adults were: *Cratoplastis diluta* Felder, *Bertholdia albipuncta* Schaus, *B. flavidorsata* Hampson, *B. specularis* H.S., and *Melese rusatta* Hy. Edw. Species of *Baritus*, *Ormetica*, and *Paraeuchaetus* accepted the diet, but none of the larvae survived beyond the 3rd instar.

Comment: The arctiids were easily reared on this diet. They moulted regularly, seldom failed to complete development and emerge properly as adults, and had a low mortality rate. The ctenuchids, on the other hand, often refused the diet, moulted erratically, and often failed to pupate or emerge properly. Normal adults were obtained only for *Gymnelia*, *Hypocharis*, and *Saurita*. The few adults that emerged for *Macrocneme* and *Poliopastea* were deformed. The addition of linolinic acid did not improve their emergence.

I suspect the oligophagous habit of ctenuchids was responsible for the erratic response to the bean-wheat germ diet. In *Macrocneme* the adverse effects of the diet usually showed up by the 5th instar. The rate of growth slowed, and a larva either failed to pupate or it lingered between instars and fed erratically. Usually it succumbed while attempting to moult or pupate. Whether these delays in development were due to nutritional deficiencies or to a change in the physical or chemical condition of the diet is unknown.

Early Stages

Information on the early stages of *Macrocneme* is based on material seen from the following species:

1. Egg: *adonis, chrysitis, coerulescens, durcata, orichalcea, thyra,* and *thyridia.*
2. Larva: *chrysitis* (1st instar), *coerulescens* (full-grown larva), *durcata* (3rd or 4th instar), *orichalcea* (exuvium only, f.g.l.), *thyra* (f.g.l.), *thyridia* (f.g.l.).
3. Pupa: *coerulescens, orichalcea,* (shell only), and *thyra.*

(For detailed descriptions of these stages see the individual species.)

Oviposition: Natural site unknown. In the laboratory, field-collected females oviposited in plastic bags, laying eggs in small clusters of 8-15 eggs arranged in short rows or occasionally singly. The greatest number of eggs laid by one individual was 45 (*M. coerulescens*, RED rearing lot 2F75). Other females deposited as many as 41 eggs (*M. coerulescens*, lot 1F75), 35 (*M. thyra*, lot 13E75) and 28 (*M. durcata*, lot 6A75). All eggs were usually laid within 24 hours and the female would survive 2-3 days subsequently.

Egg: The egg is semi-spherical, white, pale yellow or pale green, and measures 0.8-1.23 mm in diameter. It is transparent with tiny hexagonal reticulations over the surface. Approximately 24 hours before hatching, the dark mandibles of the larva become visible through the chorion and can be seen moving at least 12 hours before emergence. Eclosion occurs in 4-6 days, but may be delayed as long as 8 days.

Larva: Upon hatching, the larva eats the chorion before moving to other food. After each moult it consumes its own exuvium, leaving behind only the head capsule. The first moult is readily detected by a change in the head-capsule color (a darker brown to black) and by the appearance of secondary setae. By the third moult the basal sclerotization of various verrucae becomes strikingly iridescent green or blue. By the fourth moult, black verricules appear on abdominal A1 and A7; extra-long black tufts extend forward on T2; long, paired, white setae extend from T3 and A8; and the head capsule again changes color.

Larval development on the artificial diet was too erratic to determine with certainty the number of instars or the total development time (egg to adult). The general growth pattern parallels that for another ctenuchid, *Ceramidia butleri* [= *Antichloris viridis* Druce] which Harrison (1959) reared on its natural host (bananas). I suspect the pattern of 6-7 instars developing over three weeks and an adult emerging 10 days later is common to many ctenuchids and is representative of *Macrocneme*. Apparently the number of instars and the onset of pupation are not rigidly controlled in ctenuchids. Harrison found all *Ceramidia* larvae to pass through 6 instars, and 90% to continue on as 7th instars. *Macrocneme* larvae initiated pupation from 5th instar as well as from 6th and 7th instars. Some moulted additionally to 8th and 9th instars but never pupated. Whether the inadequacey of the artificial diet forced premature pupation in the 5th instar or delayed it indefinitely in 8th and 9th instars remains unclear. The first occurrence of a mature larva does not necessarily signal the final instar. I consider the 5th instar larvae of *thyridia, coerulescens,* and *thyra* to be mature, since the fourth moult (instar 5) produces larvae with a dramatically different appearance (see above) that remains unchanged in all subsequent moults. I cannot distinguish the instar number after the fourth moult except by slight differences in the head-capsule size. The variation in the number of moults makes it difficult to predict the final instar. Without additional data, it is not clear which instar is preferred for pupation, i.e., is the final instar. In *Ceramidia* there are apparently two "final" instars, with the 7th preferred. The same is probably true for *Macrocneme.*

Pupa: Before pupating the larva spins a flimsy, semi-transparent cocoon with off-white to light tan silk. The larva methodically twists back on itself during this process to detach the black (and sometimes white) plumose setae from the abdominal verrucae, and incorporates them into the silk. According to Mabilde (1896), pupation takes place among dry leaves and stems. The suspended pupa measures 13-16 mm in length and is visible through the semi-transparent cocoon. When fresh, the pupa is bright brownish yellow, with the appendages and abdominal segments outlined in black. The aposematic resemblance to Hymenoptera is striking (see Figs. 193-194). The color darkens before emergence, but the black banding remains apparent. Bernardi (1973) reports similarly colored pupae in *Saurita* and *Metaloba,* and suggests they are unpalatable and warningly-colored as a mechanism to ward off predators. The duration of the pupal stage is probably 10 days (see *M. leucostigma*), although my laboratory-reared specimen of *M. coerulescens* emerged in 12.

Adult

Seasonal Occurrence: Collection records show adults are present in every month of the year. Fig. 3 illustrates the abundance of species fluctuating by month. I suspect this variation is influenced by the relative abundance of the host-plant during wet and dry seasons, since the tropical rain and semi-evergreen forests in which *Macrocneme* occur are, on the average, drier between June and October than between December and May (Eyre, 1968). The relative availability of *Macrocneme* corresponds to these periods, with the largest number of species (23) flying in December and the fewest (14) during June. Whether this correlation is a seasonal response to food-plant availability is unknown. Knowledge of the host-plant should eventually clarify the mechanism.

Generations: While there are no data available on the number of generations per year for *Macrocneme,* the frequency of adults throughout the year suggests that multivoltinism is universal in the genus. *M. imbellis* and *melanopeza* are known from too few specimens to be able to predict multiple generations. Dates for *ferrea* and *habroceladon* are scant, but adults appear over a span of four months, suggesting that more than a single generation occurs. Of the remaining 25 species, all occur in six

Table 2. Sex Frequencies in 30 species of *Macrocneme*

SPECIES	FREQUENCY			
		Sex Ratio		
	Total Seen	% Male	% Female	M : F
adonis	452	58	42	3:2
lades	920	57	43	3:2
thyra	565	62	38	3:2
chrysitis	333	41	59	2:3
iole	190	38	62	2:3
cabimensis	62	43	57	2:3
durcata	146	38	62	2:3
orichalcea	45	49	51	1:1
aurifera	53	49	51	1:1
thyridia	117	57	43	3:2
coerulescens	248	83	17	4:1
semiviridis	62	58	42	3:2
oponiensis	26	62	38	3:2
ancaverdia	31	55	45	1:1
immanis	64	72	28	3:1
habroceladon	17	100	**	**
tarsispecca	22	91	9	10:1
melanopeza	2	100	**	**
zongonata	36	75	25	3:1
imbellis	1	100	**	**
bodoquero	26	77	23	3:1
leucostigma	670	51	49	1:1
bestia	28	46	54	1:1
cyanea	30	53	47	1:1
cupreipennis	44	73	27	3:1
ferrea	4	75	25	3:1
megacybe	13	77	23	3:1
mormo	47	74	26	3:1
pelotas	52	73	27	3:1
maja	***			
Total	4326	57	43	3:2

** Data not available, females unknown.
*** Identity uncertain; see text discussion.

or more months of the year, records for 17 of these (68%) appear in at least nine months and 7 (28%) are found year-round. The generation time is probably similar to that for *Ceramidia* i.e., 4-6 weeks depending on the season and availability of food (Cevallos, 1957; Harrison, 1959). Even when reared on an inadequate artificial diet, one female of *M. coerulescens* (RED lot 8K74) emerged within 60 days, indicating that at least six generations are possible within a year even under adverse growing conditions. Whether this number is affected by a diapause, especially in the drier semi-evergreen and deciduous forests and the cooler montane forests is unknown. Given the year-round availability of most of the species I suspect the life cycle does not include a diapause.

Active Period: As many as 8 species may fly together, though not in equal proportions. One species usually is commonest and the others comparatively scarce, occurring only intermittently. Of 81 examples of *Macrocneme* collected over a period of a week in May at Guatopo National Park, Venezuela, 64% were *M. thyra* (17 males, 35 females) and the remainder were divided as follows: 12.3% *adonis* (3,7), 12.3% *durcata* (3,7), 2.5% *coerulescens* (2,0), 5% *aurifera* (1,3), 1.2% *lades* (1,0), 1.2% *semiviridis* (0,1), and 1.2% *thyridia* (0,1). Among these species, appearances are so deceptive that the presence of an uncommon species is easily masked. Failing to detect a species is especially true when males are not available. The synonymy of *spinivalva* with *aurifera* illustrates the taxonomic problem created when males are not available. The occurrence of *thyridia* in Guatopo was discovered only because the larva was different. No males were available, and the single female resembled a small version of *thyra*. Only when the genitalia was examined did the association with *thyridia* become evident.

Sex Ratio: Collections of adults often show a sexual bias. On an individual basis, 16 species show some degree of male bias, while 4 (*chrysitis, iole, cabimensis, durcata*) favor the female, and 6 (*ancaverdia, aurifera, bestia, cyanea, leucostigma, orichalcea*) are essentially unbiased. Three species (*habroceladon, imbellis, melanopeza*) are obviously male-biased, as we lack any information about females.

Among the abundant species (< 100 specimens) a 3:2 sexual ratio is usually maintained whether the bias favors males or females. Only in *M. coerulescens* and *M. leucostigma* is the ratio appreciably different. In *M. leucostigma* the unbiased 1:1 ratio is not surprising considering the large sample size (n=670). In *M. coerulescens* the high male-bias (4:1) is probably characteristic of the species, and supports my own experience that *coerulescens* females are particularly scarce at light. The same may be true in Trinidad for *M. thyra* where Beebe and Kenedy cite the species as highly male-biased. Out of 326 individuals collected, only 10 (3%) were female. The difference in ratio from that given in Table 2 is not readily explained. I have never seen other species as strongly biased.

Among the less abundant species (> 100 specimens) showing a male bias, the ratio of males to females varies from 3:1 to 4:1. These are higher than the average for the genus, probably because males are easier to identify than females. In the case of *M. tarsispecca*, the 10:1 ratio results from our uncertainty about the female's identity. I suspect that as the females become better known most of these species will eventually show only a slight male-bias or prove to occur inequal numbers. A male bias in light-collected Lepidoptera is not uncommon and is probably an artifact of the collecting method.

Curiously, the three Central American species (*chrysitis, iole,* and *cabimensis*) appear to be linked by a female-collection bias. I am uncertain whether the common occurrence of a female bias is merely coincidental or of phyletic significance. These three species are the most closely related by genitalia in the genus; hence a common link in their activity behavior is likely.

Migrations: On occasion, emergence in *Macrocneme* is synchronized. Single species have been observed in large numbers at light. This behavior is reported for *M. chrysitis* and *M. iole*. During the day adults usually fly alone, but Beebe and Fleming (1951) have recorded a gathering of 24 individuals (species unknown) migrating through Portachuelo Pass in Venezuela. These migrations are infrequent and unpredictable, and I doubt that they are an inherent part of *Macrocneme* behavior. The diurnal gregariousness was probably a chance occurrence resulting from one species emerging at the same time that conditions developed for a migration. Two other species (*M. coerulescens* and *M.*[?] *yepezi*) were seen at the same time in low numbers (n = 1, n = 3 respectively).

Defense

Plant Attraction and Palatability: Many ctenuchids are strongly attracted to plants containing pyrrolizidine alkaloids (PA's) and cardiac glycosides (CG's) (Rothschild, von Euw, and Richstein, 1973; Pliske, 1975). These secondary plant substances are distasteful to predators and highly poisonous. While the borages and composites listed in Table 1 contain these toxins, the infrequency of visits by adult *Macrocneme* suggests that the attraction is unrelated to their chemical defenses. *Heliotropium* is a PA-containing borage that has often been used to attract ctenuchids (Hagmann, 1938; Moss, 1947; Beebe, 1955), yet Beebe and Kenedy (1957) report that *Macrocneme* is not attracted, and Pliske (1975) records only infrequent visits by *M. adonis* and *M. lades*. The function of the attraction in ctenuchids is not understood. Rothschild et al. (1973) suggest that both sexual attraction and defense are involved.

There is no evidence that toxins play a role in sexual attraction in *Macrocneme*, though they certainly must in other genera. I have previously separated *Poliopastea* from *Macrocneme* by their possession of a large ventral pocket on the abdomen (Dietz & Duckworth, 1976). The pocket is full of hair scales which I presume are involved in the mating process. The finding by Pliske (1975) that only males of *Poliopastea* visited the PA-containing plant *Heliotropium* suggests that they, like danaids (Boppré, 1978), are obliged to visit such plants as a means of acquiring necessary precursors for pheromones used during mating.

If there is a benefit to *Macrocneme* from the ingestion of plant toxins, it is probably as a predator deterrent. As Rothschild, Aplin, et al. (1979) suggest for arctiids, the larvae of *Macrocneme* may be opportunists, storing PA's if they occur in food plants encountered during the larva's perambulations. The known host-plants for *Macrocneme* contain both PA's (*Mikania*) and CG's (*Lantana, Mesechites*) (see table in Rothschild et al., 1973.) The pairing of these two toxins agrees with Rothschild's observation (1973) that some ctenuchids seem to favor "pairs of toxins."

The significance of this pattern is not known, although ease in storage is suggested. The outcome from their ingestion is a distastefulness that protects the larva and presumably the adult.

We now know three host-plants for *Macrocneme*. Each is from an unrelated family, i.e., *Mesechites* (Apocynaceae), *Mikania* (Compositae), and *Lantana* (Verbenaceae). This host-plant relationship conforms to a pattern suggested by Rothschild (1973) for aposematic insects that sequester toxins. Here, insects (even from the same subfamily) will select isolated genera of plants from unrelated families sharing the same secondary plant substances. In effect, their host-plant selection establishes a Müllerian mimicry complex where all the species become unpalatable models. Such an association between *Macrocneme* and some species of *Poliopastea* is suggested by the recent discovery that larvae of both genera feed on the same species of Apocynaceae. Janzen (unpub.) has reared two species of *Poliopastea* on *Mesechites trifida*. The adults are deceptively similar to each other and have often been confused with *Macrocneme* species. In one (82-SRNP-615 [UPP]) the emergent male has white-tipped hind tarsi, and in the other (84-SRNP-372 [UPP]) the hind tarsi are black. These minor differences often indicate distinct taxa.

Predators vary in their response to the distastefulness of *Macrocneme* adults. In experiments by Beebe and Kenedy (1957), ants avoid freshly dead specimens, but frogs, lizards, and mantids will occasionally eat them. *Cebus* monkeys will reject them (Blest, 1964) and birds will assiduously avoid them even to the point of an escape response (Beebe and Kenedy, 1957).

Mimicry: My observations confirm those of Kaye (1913), Beebe and Kenedy (1957), and Blest (1964) that in both facies and behavior, adult *Macrocneme* are excellent mimics of pompilid wasps. They may also mimic ichneumonid wasps, as suggested by Kettlewell (1959). The antennae are white-tipped and the base of the abdomen is white-spotted, giving the appearance of a constricted waist. The wings and abdomen are patterned with blue and green iridescent scales. The hind legs are long and heavily scaled (hence the name: *macro* = large, *cneme* = leg). The legs are carried downward and backward in flight. An adult flies slowly in a direct line, wasplike, at the edge of forests, and along roadsides and streams, in bright sunshine. Beebe and Kenedy noted that captured adults, when released, will spiral upward fairly slowly or circle several times before flying away. When they alight, the wings are held back at a 20° angle and flat. The hindwings are overlapped with only the inner margin visible. At the moment the moth settles the wings often vibrate slightly, and the antennae will wave momentarily. During walking, the movement is slightly jerky and the wings are held up and back at a 30° to 40° angle.

Sound Production: Both sexes in every species possess tymbal organs on the metepisternum (e.g., see Figs. 197 and 200). As in other ctenuchids, these presumably function to produce sound (Blest, 1964). An unscaled area along the fore edge of the sclerite is modified with a row of minute, transverse corrugations forming microtymbals. Each groove contains a single hair-scale, the function of which is not known (See Figs. 195-196 and 198-199 for differences in size). The average length of these scales appears to vary depending on the species. Their presence in every groove is in contrast to the condition found in arctiids, where only a few of the grooves are occupied (Watson, 1975). When the microtymbals (Figs. 195-200) are deformed by a contraction of the thoracic flight muscles, they buckle to produce a series of ultrasonic clicks (Blest, Collett, and Pye, 1963). The exact function of these signals is not clear. Dunning (1968) has suggested that they are aposematic displays which warn of unpalatability to bats and other potential predators. There is no evidence that these displays occur in *Macrocneme. M. adonis* and *M. thyra intacta* [sic] are the only species that have been studied. Blest (1964) found they did not produce sound when subjected to tactile stimulation and recorded bat cries. Conceivably the tymbal organ is no longer functional in *Macrocneme*. Many ctenuchids appear to be switching from a totally nocturnal existence to one that is partially diurnal. In this switch they have encountered a new predator, i.e., birds, and have had to evolve a visual mechanism (mimicry) to survive. It is possible that with a change in life-style and predator type the nonvisual aposematic mechanism of sound has become obsolete. The tymbal organ would then be a vestigial organ with no residual function.

Scent-Distributing Organs: All males have a bilobed pocket between the seventh and eighth segments of the abdominal venter (Fig. 132). Field (1975) reported a similar structure in the related genera *Antichloris, Ceramidia,* and *Ceramidioides*. It presumably is an eversible gland that functions during mating. There is no information on the mechanism involved. My suspicion is that the gland operates at the same time the genitalia are extruded. As pointed out by Blest (1964), *Macrocneme* males are often seen with their entire genital apparatus extended. When the structure is fresh it is bright yellow to yellow-brown, leading Blest to suggest that it functions as a warning signal. Instead, this behavior may be a "calling mechanism" involving the distribution of a sex pheromone or possibly is a sensory receptive organ for the female pheromones as is known in other Lepidoptera. The male genitalia are comparatively large and heavy. When the genitalia are extended, the seventh and eighth abdominal segments become flexible and allow the intersegmental pocket to then be everted or at least partially opened for the release of a pheromone. There is a certain amount of wing-waving during this procedure, which Blest considered was aposematic behavior. Alternatively, it may be a method to aid in dissemination of the pheromone.

GEOGRAPHICAL DISTRIBUTION

General

Like many Euchromiine ctenuchids, the genus *Macrocneme* is endemic to the New World and is restricted mostly to the tropical and subtropical regions of Latin America and Brazil in areas of high annual rainfall. The genus ranges continuously from the southwestern United States to northern Argentina and eastward to the Atlantic perimeter of Brazil. The northern and southern extremities of the range are presumably limited by intolerance to the more temperate climate, with *M. chrysitis* being the sole representative at 29° N lat. (Texas, USA), and *M. leucostigma* at 34° S lat. (Buenos Aires, Argentina). High concentrations of species are found in Colombia (14), Perú (15), and Brazil (17) (see Fig. 1 and Map 3). While these concentrations are in part reflection of the geographical size of political units, they also reflect the influence of historical changes in the creation of Quarternary refugia (see Brown, Sheppard, and Turner, 1974; Haffer, 1979; Simpson, 1979). Only one species, *M. thyra*, is recorded from the Antilles, on Grenada, where it is probably a late invader from Trinidad.

Distribution Patterns

Aside from the 3 species *M. adonis, lades,* and *thyra,* which are unusually widespread (Mexico to Brazil), the distributional patterns for individual species conform to three major geographical regions defined by K.S. Brown (1979) (For abbreviations, see Map 1):

1. Northwestern (NW): Neotropical Mexico, Central America, northwestern South America; also the interior valleys of Colombia, the Pacific coast of Colombia and Ecuador; the Caribbean coast of Colombia and Venezuela; the eastern slopes of the Andes to the Serranía La Macarena; occasionally reaching to the Guianas and into the Amazon as far as Manaus and the Rio Marañon.

Associated species (9): *M. chrysitis* (Map 6) endemic to Mexico and Guatemala, replaced by *M. iole* (Map 7) in Costa Rica and Panamá; *M. cabimensis* (Maps 9 and 10) from Belize to Ecuador, clinally variable, forms a mimetic complex with

M.semiviridis (Map 8) and *M. oponiensis* (Map 10) in Colombia and Ecuador; *M. coerulescens* (Map 17) geographically variable, predominantly in Venezuela and Colombia; *M. lades* and *M. thyra* (Maps 13-15) widely sympatric, geographically variable; *M. thyra* rare north of Panamá; *M. adonis* (Maps 4 and 5), the only sexually dimorphic species.

2. Amazonian Hyalea (AB + AN + GS): Orinoco Basin, the Tepuis, the Guianas, west to the Andes and south to Bolivia; may invade the Northwestern region in northern Venezuela, Colombia, and Darien of Panamá; may extensively invade central Brazil, occasionally reaching Espiritu Santo.

Associated species (16): *M. thyra* and *M. lades* (Maps 13-15), broadly sympatric, racially variable with brown phenotype from old Guiana Shield; *M. adonis* (Maps 4 and 5) occasional in Amazon, absent from Andes; *M. thyridia* (Map 16) prevalent in eastern Venezuela and the Guianas, occasionally south to Mato Grosso, Perú, and Bolivia; *M. aurifera, M. orichalcea* (Maps 19 and 22) in the Guianas, eastern Venezuela, Amazon Basin to Perú; *M. durcata* (Map 20) geographically variable, coastal Venezuela, eastern Andes to Bolivia; *M. ancaverdia* (Map 18) primarily Andean, occasionally from the Guianas; *M. habroceladon, M. tarsispecca, M. immanis* (Maps 11 and 12), closely allied group, mostly Andean but *M. immanis* absent from Colombia; *M. bodoquero, M. imbellis, M. zongonata* (Maps 21, 22, 23) predominantly Amazonian, with *M. bodoquero* occasionally reaching southeastern Brazil; *M. melanopeza* (Map 22) from single Peruvian record; *M. leucostigma* (Map 27) common in southern Brazil and radiating westward to Perú.

3. Atlantic (BS + AT): southeastern Brazil to Bahía, extending occasionally to Paraíba and Rio Grande do Norte, occasionally crossing northeast to Pará, extending frequently to the interior and south to the limits of climatic tolerance, or restricted solely to the subcoastal Brazilian sierras.

Associated species (12): *M. bestia, M. cupreipennis, M. ferrea, M. megacybe, M. mormo, M. pelotas* (Maps 16, 23, 24, 26) endemic to southeastern Brazil; *M. cyanea* principally southern Brazil, single record from northern Argentina (Map 25); *M. adonis, M. thyra, M. lades* (Maps 4-5, 13-15) only occasionally in southeastern Brazil; *M. bodoquero* (Map 21) principally Amazonian, with a single record from southeastern Brazil; *M. leucostigma* (Map 27) primarily southern Brazil and northern Argentina, occasionally from Perú.

Table 3 lists species according to major geographical regions. *M. durcata* and *M. orichalcea* are placed tentatively in the Atlantic region because the identification of two females of *M. durcata* and one of *M. orichalcea* from southern Brazil is uncertain. The disjunctions make me suspect the identifications are incorrect. Other problems include *M. aurifera* males, which are relatively common in Trinidad, but the species was described from Perú where males still have to be collected. *M. bodoquero* is predominantly Amazonian, but one male examined from Cantareira (São Paulo) has confirmed its presence in southeastern Brazil.

Table 3. Biogeographical Distribution in *Macrocneme*

SPECIES	MAJOR REGIONS			ASSOCIATED QUATERNARY REFUGIA [a]
	Northwestern	Amazonian Hyalea	Atlantic	
1. adonis	X →→	→→ X →→	→→ X	AB, BS, GS, NW
2. lades	X ←←	←← X →→	→→ X	AB, AN, BS, GS, NW
3. thyra	X ←←	←← X →→	→→ X	AB, AN, BS, GS, NW
4. chrysitis	X			NW
5. iole	X			NW
6. semiviridis	X			NW
7. cabimensis	X			NW
8. oponiensis	X			NW
9. coerulescens	X →→	→→ ?		NW, AN (?)
10. durcata		X →→	→→ ?	AN, NW, GS (?)
11. thyridia		X		AB, AN, GS
12. orichalcea		X →→	→→ ?	AB, AN, GS
13. aurifera		X		AN, BS, GS
14. ancaverdia		X		AB, AN, GS
15. immanis		X		AN
16. habroceladon		X		AN
17. tarsispecca		X		AN
18. melanopeza		X		AN
19. zongonata		X		AB, AN
20. imbellis		X		AB
21. bodoquero		X →→	→→ X	AB, BS
22. leucostigma		X ←←	←← X	AN, BS
23. bestia			X	BS
24. cyanea			X	BS
25. cupreipennis			X	BS
26. ferrea			X	BS
27. megacybe			X	BS
28. mormo			X	BS
29. pelotas			X	BS
30. maja			? [b]	

a. AB = Amazon Basin; AN = Andean; AT = Atlantic; BS = Brazilian Shield; GS = Guiana Shield; NW = Northwestern.
b. Species identity and type locality uncertain; see text.
Arrows = direction of radiation from suggested centers of origin.

Table 4. Sympatry of *Macrocneme* species
by political subdivisions (see Map 3)

	POLITICAL UNIT	No. of SPECIES
1.	Rio de Janeiro, Brazil	9
2.	São Paulo, Brazil	9
3.	Cundinamarca, Colombia	8
4.	Meta, Colombia	8
5.	Cochabamba, Bolivia	8
6.	Aragua, Venezuela	7
7.	Miranda, Venezuela	7
8.	Pará, Brazil	7
9.	Junín, Perú	7
10.	Puno, Perú	7
11.	San Martín, Perú	7
12.	Antioquia, Colombia	6
13.	Boyaca, Colombia	6
14.	Valle, Colombia	6
15.	La Paz, Bolivia	6
16.	Santa Cruz, Bolivia	6
17.	Amazonas, Brazil	6
18.	Santa Catarina, Brazil	6

M. coerulescens is questioned in the Amazonian Hyalean Region due to my uncertainty about the relationship of the disjunct blue Peruvian form to the polytypic(and perhaps clinal) forms of *M. coerulescens* from Venezuela and central Colombia. The appearance as a disjunct morph in Perú makes me suspect that more than races are involved here and that the disjunction is an artifact created by insufficient differences in the genitalia to accurately determine taxonomic placement.

Sympatry

Except at the range extremities, two or more *Macrocneme* species often occur together. A high degree of sympatry is the rule (Maps 2 and 3). When broad political boundaries are subdivided into smaller units (states, departments, provinces) (Fig. 2), two or more species occur in 68% of the units. Concentrations of 6-9 species in 12% ofthe units (Table 4) are partially the result of collector bias, but a pattern emerges paralleling the peaks shown in Fig. 1 for the three major geographical regions. Sympatry remains high even when the species composition changes from one region to another. An apparent lack of sympatry (32%) is perhaps due to our incomplete knowledge of species distributions. The known overlap among species should increase as the Neotropical fauna becomes better known.

Effect of Altitude

The range of elevations at which *Macrocneme* occur is defined either by vegetation type or by the topography. None of the species is presently known to be restricted to a narrow altitudinal range. Instead, the occurrence of each is probably determined by the presence of tropical rain and tropical semi-evergreen and deciduous forests (see Eyre, 1968). Their uppermost limit is the tropical montane forests of the Andean Cordillera. The genus ranges in elevation from sea level to 1800 m (*M. durcata*, Coroico, Bolivia) and to a high of 2800 m (*M. ancaverdia*, Riobamba, Ecuador). Occasionally species are recorded at low elevations within areas that are predominantly broad-leaved tree savannas. Here their presence most likely is restricted to the gallery forests. The southern Brazilian species occur mostly in the tropical forests along the coast and inland in the semi-evergreen and deciduous forests. They range in elevation from 50 m to 1400 m (*M. cyanea*). The difference in range from the Andean species is due to the absence of high montane forests. I suspect most species would tolerate higher elevations, but have become so closely tied to their present vegetation types that they are unable to cross vegetation barriers to the west where the host-plant is absent. Only *M. leucostigma* has successfully crossed both mixed evergreen forests and pampas regions to reach the higher montane forests of the Andes. A possible mechanism is that *M. leucostigma* is either polyphagous or is associated with a host-plant that is not restricted to tropical rain and semi-evergreen vegetation forests.

Table 5. Endemism in *Macrocneme* Species

Species	Northwestern	Amazonian Hyalea	Atlantic
Total	9	16	12
Endemic	6	11	7
% Endemic	67	68	58

Endemism

Most species of *Macrocneme* are restricted to one major geographical region (Table 5). While some are widespread within a given region and require careful examination for identification, others are so narrowly restricted that the locality will often automatically eliminate certain species from consideration in an identification. Seven species from southeastern Brazil are so narrowly endemic that their identification from localities beyond the coastal states would be suspect. From Tables 5 and 6 it is evident that endemism is high. Only 5 species (17%) cross to a second region, and 3 (10%) traverse all three regions. Although not endemic, these last 3 species (*M. adonis, M. lades, M. thyra*) are more concentrated in the Northwestern and Amazonian regions than in the Atlantic. Most likely they are limited in distribution more by climatic or host-plant considerations than by geological events which appear to have influenced the limits of the more narrowly endemic species (see Haffer, 1979, for summary).

Differences in size of geographical area obscure the fact that the more narrowly endemic species occur in about equal numbers in the three major regions. Southern Brazil has the most narrowly endemic group with 8 species, 7 from the coastal states and one (*M. leucostigma*) that has radiated extensively into the interior. In the Amazonian Hyalea region, the most narrowly endemic species are either strictly Andean (*M. habroceladon, M. tarsispecca, M. immanis, M. melanopeza*) or Andean with an Amazonian element (*M. imbellis, M. zongonata*). In the Northwestern region, *M. iole* replaces *M. chrysitis* in Central America, and where *M. semiviridis, M. oponiensis,* and *M. cabimensis* are narrowly restricted in Colombia and Ecuador, only *M. semiviridis* extends eastward to Venezuela, and *M. cabimensis* northward to Belize.

Table 6. Relationship of *Macrocneme* Species among Major Geographical Regions

Species		Northwestern	Amazonian Hyalea	Atlantic
Northwestern	a	–	3	3
	b	–	17	13
	c	–	33	33
Amazonian Hyalea	a		–	5
	b		–	18
	c		–	42
Atlantic	a			–
	b			–
	c			–

a = Species common to both; b = Total species not common to both; c = Percent relationship. Percent calculated by dividing "a" by the smaller total of endemic species (from Table 4) between two regions to correct for difference in faunistic size.

Endemism on a broader scale can be illustrated by comparing pairs of major geographical regions for the number of species held in common and separately. As one might expect, the percent relationship decreases the more distant the regions are from each other (Table 6). The Northwestern region has the same relationship (33%) to both the Amazonian Hyalea and the Atlantic, due to their sharing the same 3 species. The higher relationship (42%) between the Amazonian Hyalea and the Atlantic region is in part a result of their contiguous border, and they share 2 additional species, *M. bodoquero* and *M. leucostigma*. If the questionable records for *M. orichalcea* and *M. durcata* are included, the relationship is 58%, which is not a surprising faunistic affinity between contiguous areas.

Distribution Centers

An analysis of endemism suggests that *Macrocneme* has radiated from certain centers. Maps 2 and 3 illustrate species concentrations by political subregions. These areas correspond remarkably well to the Quarternary refugia as proposed by various authors (see K.S. Brown, 1977 and 1979; Haffer, 1979; Simpson, 1979). These refugia are particularly useful when discussing races. Brown has defined 44 geographic centers of endemism. Since I have treated *Macrocneme* only at the species level, I have not attempted to associate each species into one or more of these narrowly defined centers of endemism. Rather, I have defined five refugia which correspond to species abundances, and which in large part parallel the limits of the major geographical regions. The principal exception is the Amazonian Hyalea, which I have divided into three sub-regions to better reflect geological events such as the Tertiary age of the shield areas and the uplifting of the Andes. Following are the refugia (Map 1):

1. Northwestern (NW): Neotropical Mexico, Central America, and northwestern South America.
2. Amazon Basin (AB): drainage system of the Upper and Lower Amazon valleys, the true Hyalea without the Orinoco.
3. Andean (AN): eastern slopes of the Andes from coastal Venezuela to Argentina; it differs from AB principally in elevation.
4. Brazilian Shield (BS): south-central highlands of Brazil, including the Atlantic region (AT).
5. Guiana Shield (GS): Orinoco Basin, including the Guiana Highlands and the plateaus west of the Tepuis.

In Table 3 I have assigned these refugia on the basis of a species' presence rather than strictly by region of greatest abundance. This method obscures the center of distribution when more than one region is involved. Where there is overlap, one region always dominates and is probably the major center of distribution. For example, *M. leucostigma* is predominantly BS but has radiated westward to AN. *M. bodoquero* is strongly associated with AB and only weakly with BS, due to a single record from São Paulo. *M. thyridia* and *M. aurifera* are most strongly associated with GS and weakly with both AB and AN. *M. orichalcea* and *M. zongonata* are most strongly affiliated with AB, but have weak elements in AN. *M. coerulescens* is undoubtedly tied to NW and only weakly so to AN if the blue Peruvian form is conspecific. *M. durcata* and *M. ancaverdia* are strongly AN and weakly represented in GS.

I would speculate that *M. adonis* originated in Central America, where it is most commonly collected, and then radiated north to Mexico to the limits of climatic tolerance and south into continental South America, where it has become an increasingly less dominant form in comparison to its congeners the further it is removed from its center of origin. Its presence in Pernambuco and Bahía is based on single male records. *M. lades* and *M. thyra* appear to have originated in the

Guiana Highlands and to have radiated westward into the Transandean (i.e., Northwestern) region. In response to a different climatic regime they have undergone similar changes in appearance and remained sympatric except north of Panamá, where *M. thyra* becomes scarce.

Parallel Geographical Variation

Clarification of taxonomic relationships has revealed that parallel geographical variation is exhibited by various species pairs and triplets. The underlying cause most likely is Müllerian mimicry, because species in these complexes share a high similarity in external appearance and are largely sympatric. Traditionally, Müllerian mimicry is supposed to function most efficiently when a great large number of individuals resemble a single phenotype. Polymorphism is considered a disadvantage, since the number of individuals protected by a single phenotype is reduced, as predators must learn to associate distastefulness with more than one form. Nevertheless, in *Macrocneme* all species are variously polymorphic and most remarkable are the mimetic complexes involving pairs and triplets which vary in parallel. Similar variation is well known in other groups, most notably heliconiid butterflies where in the genus *Heliconius* alone there are at least 30 morphs for two widespread species, *H. melpomene* and *H. erato* (Emsley, 1964; Turner, 1971, 1977; K.S. Brown, 1979). In North America Doyen and Somerby (1974) found tenebrionid beetles in the genera *Coelocnemis* and *Eleodes* exhibiting parallel polymorphism in 16 species and geographic races.

In *Macrocneme*, parallel variation exists in the color and pattern of wing iridescence. Color changes are most obvious for four species occurring sympatrically in Colombia and Ecuador. There are two forms, one blue to blue-green and the other green to yellow-green. In Andean Ecuador (Paramba) *M. cabimensis*, *oponiensis*, and *thyra* occur as a yellow-green triplet, and in Santo Domingo de los Colorados *oponiensis* is found paired with yellow-green *semiviridis*. In western Colombia, *semiviridis* occurs in a blue-green form with *cabimensis* (Nare; Cananche) and as a green form with *thyra* (Rio Dagua; Honda). In north-central Colombia (Río Opón), *oponiensis* and *semiviridis* are again paired, but are blue-green. While I have not seen any pairs of *cabimensis* and *oponiensis* from identical localities in Colombia, their pairing with *semiviridis* in various localities – e.g., Nare, Muzo, Cananche – suggests that all three will be found together as more material becomes available. Likewise, *semiviridis* is likely to be found in association with the yellow-green forms of *cabimensis* and *thyra* in Ecuador.

Variation in wing pattern among species pairs is best exemplified where *lades* occurs with *thyra* in the Guianas and with *cabimensis* in Central America. In the Guianas, the phenotype for *lades* mimics that of *thyra* in having a distinctly brown ground-color and a brassy sheen to the iridescence. Elsewhere in their largely sympatric ranges the ground color is darker brownish black and the color of the iridescence varies in parallel depending on the population, being blue for both species in Guatopo, Venezuela, and a mixture of blue and green in Choroní. Other parallels are discussed for these species under the individual species treatments in the text.

When *cabimensis* ceases to be associated with *oponiensis* and *semiviridis* in Colombia, it loses its resemblance to these species and assumes an appearance much like *lades* when the two are sympatric in Central America. (See *cabimensis* for further discussion.)

The formation of pairs and triplets varying in parallel may relate to the geological history of the region. Turner (1971) has suggested for the *melpomene-erato* complex that the effect of the glacial periods was to isolate populations, creating an environment in which local differentiation was initiated. Subsequent spread of these races during the interglacial periods resulted possibly in polymorphic hybrid areas in which related species, driven by the selective forces of mimicry, came to resemble each other. This mechanism would appear plausible for explaining similar paired variation in *Macrocneme.*

KEY TO SPECIES

(Based on the Male Genitalia)

1a.   Juxta extending beyond base of ventral processes of valvae (Fig. 5) 6

1b.   Juxta approx. level with or not extending beyond base of ventral processes of valvae 2

2a.   Juxta with double-pronged margin (Fig. 60) .... **durcata** n. sp.

2b.   Juxta reduced to thin sclerotized rod (Fig. 73) .... **aurifera** Hampson

2c.   Juxta not as above 3

3a.   Juxta with spiny patch in membrane at left 4

3b.   Juxta without spiny patch 5

4a.   Left spine of aedeagus strongly hooked sinistrad at tip (Fig. 106) ....
.... **bestia** n. sp.

4b.   Left spine of aedeagus not hooked (Fig. 57) .... **thyra** Möschler

5a.   Dorsum of uncus appearing inflated, lateral margins asymmetrically flanged (Fig. 121) .... **megacybe** n. sp.

5b.   Dorsum of uncus not inflated, lateral margins symmetrially flanged (Fig. 36) .... **immanis** Hampson

5c.   Dorsum of uncus narrow, lateral margins not flanged (Fig. 93) ....
.... **zongonata** n. sp.

6a.   Aedeagus with 1 to 3 spines present 7

6b.   Aedeagus without spines (Fig. 26) .... **semiviridis** Druce

7a.   Aedeagus with one dorsal spine 8

7b.   Aedeagus with two dorsal spines 11

7c.   Aedeagus with three dorsal spines (Fig. 70) .... **thyridia** Hampson

8a.   Spine of aedeagus at left margin of dorsum 9

8b.   Spine of aedeagus at right margin of dorsum 10

9a.   Aedeagus spine small, spatulate process present (Figs. 20a-b) .... **iole** Druce
9b.   Aedeagus spine prominent, spatulate process absent (Fig. 16) ....
.... **chrysitis** (Guérin-Méneville)

10a. Dorsal process of valva uniquely bifurcate (Fig. 6), spine of aedeagus strong,
directed sinistrad (Fig. 9) .... **adonis** Druce
10b. Dorsal process of valva distinctly concave at tip (Fig. 27a), spine of aedeagus
small .... **cabimensis** Dyar

11a. Dorsal processes (claspers) of valvae asymmetrical 12
11b. Dorsal processes of valvae mostly symmetrical 18
11c. Dorsal processes symmetrical, but tips long and upturned (Fig. 39a) ....
.... **tarsispecca** n. sp.

12a. Tips of claspers acuminate, inner margin of left arm with spine-like process
(Fig. 79) .... **orichalcea** n. sp.
12b. Tips of claspers acuminate, but inner margins simple (Fig. 75a) ....
.... **ancaverdia** n. sp.
12c. Tips of claspers club-like or round, left arm longer than curved right arm .. 13

13a. Uncus symmetrical, round, without lateral flanged margins (Fig. 113) ....
.... **pelotas** n. sp.
13b. Uncus skewed to left when viewed dorsally, lat. margins asymmetrically
flanged 14

14a. Juxta with spinose patch on inner membrane 15
14b. Juxta without spinose patch 16

15a. Lateral margins of uncus extending from narrow dorsum as subequal, convex
flanges, the left distad to the right; apical margin of juxta incised at right .....
(Figs. 125-126) .... **ferrea** Butler
15b. Lateral margins of uncus as unequal vertical flanges; apical margin of juxta
not incised, fully spined (Figs. 109-110) .... **cyanea** (Butler)

16a. Dorsum of uncus with 2 vertical flanges; juxta with thumb-like flap at right
(Figs. 101-102) .... **leucostigma** (Perty)
16b. Dorsum of uncus with small, center flange; juxta with round margin at right
(Figs. 117-118) .... **cupreipennis** Walker
16c. Vertical flanges absent from dorsum of uncus 17

17a. Dorsum of uncus round; margins of juxta fully spined; spines of aedeagus
small (Figs. 44-46) .... **habroceladon** n. sp.
17b. Dorsum of uncus skewed strongly to left; juxta incised at right, flap absent;
right spine of aedeagus prominently S-shaped (Figs. 32-34) **oponiensis** n. sp.

18a. Left spine of aedeagus large, pointed sinistrad 20
18b. Left spine of aedeagus approximately same size as right spine, both directed
dextrad 19

19a.   Juxta with 6-7 spines on narrow apical margin; left aedeagal spine stout, surface spinulate; right spine stubby, surface smooth (Figs. 86-87) ....
.... **bodoquero** n. sp.
19b.   Juxta heavily spinose on broad apical margin, right margin incised; left aedeagal spine with acuminate tip, right spine with blunt tip ....
(Figs. 90-91) .... **imbellis** n. sp.
19c.   Juxta spinose on broad apical margin, right margin extended into flap with margin spines pointing in opposite direction to those on left; tips of aedeagal spines similar (Fig. 98) .... **melanopeza** n. sp.

20a.   Left spine of aedeagus 4 times length of right spine (Fig. 131) **mormo** n. sp.
20b.   Left and right spines of aedeagus approximately same length 21

21a.   Margin of juxta incised on right, apex distinctly extended at left, heavily spinose (Fig. 64) .... **coerulescens** Dognin
21b.   Margin of juxta only slightly incised on right, a non-spinose flap sometimes present, apical margin with patch of spines at left (Figs. 49-52a) ....
.... **lades** (Cramer)

SYSTEMATIC TREATMENT

Macrocneme maja (Fabricius)

Zygaena maja Fabricius, 1787:106
Macrocneme maja.- Hübner (in error), 1818:15, pl.[12], figs. 65 and 66 [cited without
    reference to Fabricius as author; rectified in following].- *ibid.*, 1819:124.- Walker,
    1856:1632.- Felder, 1862:232.- Butler, 1876:371.- Kirby, 1892:128.- Hampson,
    1898:322.- Draudt, 1916:104
Euchromia (Macrocneme) maja.- Walker, 1854:248
Copaena maja.- Herrich-Schäffer, 1856:23

This is the type species of *Macrocneme*. Due to the existence of a type which I
have been unable to identify, I am treating the name as an indeterminant species
with the view that diagnostic characters may eventually be found that will resolve the
problem as to the biological entity to which the name applies.

Zygaena maja was described by Fabricius from the Hunter Collection. The
collection still exists, housed in the Department of Zoology, University of Glasgow,
Scotland. In an account of the collection's contents, Kerr (1910) listed two syntypes
for *Zygaena maja*. Through the kind intercession of Allan Watson at the British
Museum (BMNH), these syntypes were examined and photographed for me. One is
indisputably the *maja* described by Fabricius; the other appears to be a species of
Pericopid. This discrepancy is not explained – perhaps it is an error in cataloguing.
The true *maja* is a female that might be identifiable if the structure of the sterigma,
especially the seventh abdominal sternite, had been available. Unfortunately, the
genitalia no longer exist.

The type locality is often useful in identifying species of *Macrocneme*, but *Z. maja*
was recorded only as originating in "America." When Hübner proposed *Macrocneme*
in 1818, he included *maja* without a reference to its author and gave Brazil as the
type locality. His omission of Fabricius' name was obviously an oversight, which he
subsequently corrected in 1819. Apparently he was aware that Fabricius had Brazilian
material, confirming a similar observation by Forbes (1939). It is likely that the
specimen did in fact come from Brazil given that Fabricius described other
Lepidoptera from Brazil. Unfortunately, the appearance of the *maja* type is
indistinguishable from other *Macrocneme* occurring in southern Brazil, (e.g., *mormo,
ferrea*, or *megacybe*). Since I have separated these species by differences in the
genitalia, the absence of such in the *maja* type leaves the species' identity in limbo.

Macrocneme adonis Druce

(Figs. 6-9a, 146-147, 197, 199, 201-202; Maps 4 and 5)

Macrocneme adonis Druce, 1884:48, pl. VI, fig. 16
Macrocneme cinyras Schaus, 1889, 88 [NEW SYNONOMY]
Macrocneme adonis ab. *chiriquicola* Strand, 1917:84   [infrasubspecific name]

This species is unique in *Macrocneme* by virtue of its sexual dimorphism. The females have a bright blue iridescence extending over the outer half of the forewing, while the males seldom share the same intense blue, and the iridescence never extends into the apical and terminal areas. Both sexes always possess a metallic green streak on the inner margin of the forewing, and the hind legs are covered with long black scales. The uniquely bifurcate claspers and the large juxta, overlying the uncus (easily seen in dried specimens), will identify males, while the simple unmodified seventh sternite and the tiny sclerite of the lamella postvaginalis will distinguish females.

MALE. *Head*: black, vertex with scattered metallic blue, labial palpi upturned, not reaching base of antennae. *Thorax*: black, disc iridescent green to blue-green; patagia with metallic scales as small spots adjoining lateral white spots; tegulae with mesal metallic streaks, white scales absent on underside; pectus black with white spot on metepisternum, white absent on propleuron; legs black with white spots on fore and mid coxae and all trochanters; tips of hind tarsi black. *Forewing*: black to brownish black ground-color with iridescent green to blue-green fascia in median and postmedian areas; subterminal, terminal, and apical areas black; prominent black streak from basal angle characteristically dividing metallic green of inner margin from blue-green spot below cell; underside black with blue to green iridescence along costa basally, streaks sometimes extending below cell; overlapping area of inner margin gray; retinaculum black. *Hindwing*: black to brownish black, costa gray where wings overlap; iridescence restricted to small spot in limbal area or entirely absent; white scales at basal articulation; underside with metallic blue along costa, remaining surface black. *Abdomen*: dark green, iridescence dull, suffused; thin, shiny mid-dorsal and sublateral streaks obsolescing medially; mesal series of white spots on venter extending to fifth segment. *Genitalia*: as in Figs. 6-9a (drawn from RED prep. 39189, USNM; 5n); valvae unique, dorsal processes (claspers) widely bifurcate, caudal branch secondarily divided at tip; ventral process greater than twice length of clasper process; uncus narrow, skewed to left when viewed dorsally; juxta an enormous process, broad basally, narrowing neck-like medially, reaching to dorsum of uncus, apex curved inward, margin spined; dorsum of aedeagus with prominent spine on right directed sinistrad.

FEMALE. Essentially as described for male except for different wing pattern. *Forewing*: median, subterminal, and apical areas with intense blue iridescence, inner margin and basal spot below cell metallic green, separated by prominent oblique black streak to mid-cell, green sometimes suffusing into discal area; underside iridescent blue basally, brownish black apically. *Hindwing*: black along lower edge of cell, brownish black in anal area, intense metallic blue in limbal area; underside brownish black with thin green streak basally along costa and upper margin of cell. *Genitalia*: as in Figs. 146-147 (drawn from RED prep. 39187, LACM; 3n); sternite VII of sterigma unmodified, lateral margins forming narrow symmetrical flaps; intersegmental cuticula between VI and VII unmodified; inner margin of lamella antevaginalis smooth; sclerite of lamella postvaginalis scutiform, unattached to VIII; sternite VIII lacking medial protuberance, emarginate medially; dorsal wall of ductus bursae with blind pocket at left; ductus seminalis originating ventrad to base of accessory bursa; signa as for genus.

VARIATION. Length of forewing: males, 16.0-18.0 mm; females, 16.0-19.5 mm. Wing iridescence in males varies between green and blue-green, neither intensely blue as in females nor extending beyond postmedian area; occasional Amazonian specimens of *adonis* females display a mixture of green and the usual intense blue, or they can be silvery blue; color often coincides with that found in sympatric *M. orichalcea, zongonata*, or *melanopeza*; a green streak on inner margin of forewing, iridescent scales reaching the apex, and black hind tarsi will usually distinguish *adonis* females from the above congeners.

TYPE DATA. ! *adonis* Druce: female lectotype, **by present designation**, Volcán de Chiriquí, Panamá, 2-3000 ft (Champion), BMNH. ! *cinyras* Schaus: male holotype, Coatepec, Mexico, USNM. ! ab. *chiriquicola* Strand: male holotype, Volcán de Chiriquí, Panamá, 25-4000 ft (Champion), BMNH.

Druce did not indicate the number of syntypes in his original description of *adonis*. In selecting a lectotype, I have followed Hampson's (1898) use of the word "type" after Chiriquí, Panamá, when choosing a specimen for this species.
Hampson's "Ab. 1" is based on a specimen also from Volcán de Chiriquí, but the elevation [from the label] reads "25-4000 ft." It is an excellent example of a male *adonis*. Unfortunately, Strand applied the name *chiriquicola* to it, probably without ever seeing the specimen.

BIOLOGY. Occasionally adults are collected at flowers. Pliske (1975) reports occasional males attracted to *Heliotropium indicum* (Boraginaceae). Attempts by me to have field-collected females oviposit on leaves of this plant in the laboratory were unsuccessful. Presentation of leaf samples from species of *Echites, Calotropis, Senecio, Solanum, Inga, Cassia*, and *Longocarpus* similarly failed to stimulate oviposition. Females of *adonis* did not readily oviposit when confined in plastic bags. They fed to engorgement on a weak solution of honey water and survived 7-10 days usually without ovipositing. A subsequent examination of the abdomens showed the oviducts to contain fully developed eggs. Only one female from Rancho Grande (RED lot 8J74) oviposited, laying 7 pale green eggs on the 8th day of confinement.

GEOGRAPHICAL DISTRIBUTION. Widespread. Most commonly collected in southern Mexico, Central America, and northern Colombia and Venezuela; less commonly in the Guianas and the lower Amazon Basin; rarely in southern Brazil or along the Atlantic coast.

FLIGHT PERIOD. Adults are relatively common at lights and appear to fly throughout the night, but are not crepuscular. The species appears to be active throughout the year. Adult collection records are available for every month of the year, especially at sites like Rancho Grande, Venezuela, where there had been continuous collecting for many years.

REMARKS. The types of *cinyras* Schaus and *adonis* Druce are conspecific. The blue form (*adonis*) is always female and the green form (*cinyras*) is male. Hampson placed specimens of both forms under the name *adonis* in the BMNH collection, but never published the synonymy. Forbes (1939) was uncertain about the association and mistakenly maintained that the tips of the hind tarsi are white in *adonis* males. The hind tarsi in this species are always entirely black. Forbes most likely was confused by specimens of *M. lades* or *cabimensis*, which have white hind tarsi and occur sympatrically with *adonis* in Panamá.

The recent record from Mato Grosso of an *adonis* male suggests that the species is gradually spreading southward in Brazil. It is never common in the Amazon Basin, but if Brown and Mielke's estimation (1972) that the Chapada dos Guimarães is a "northwestern blend zone of the cerrado fauna with the upper Amazonian/Bolivian fauna," then *adonis* will eventually be found in the Andes of Perú and Bolivia.

M. zongonata and *orichalcea* are extremely similar to *adonis* females, but *zongonata* lacks the green streak on the inner margin of the forewing and its hind tarsi are white-tipped; *orichalcea* has two black streaks on the forewing (rather than one) and the iridescence is bluish green rather than an intense blue.

SPECIMENS EXAMINED (452): 261 males, 191 females. **MEXICO**: *Chiapas*: Cardenia, CNC; Cumbre de Arriaga, LACM; La Trinitaria, CNC; Ocozocoautla, 2700', CAS; San Cristobal de Casas, CNC; Tuxtla Gutierrez, UCB; Unión Juarez, LACM. *Colima*: Volcán Colima, SMM. *Hidalgo*: Guerrero Mill, 9000', USNM. *Oaxaca*: Juquila Mixes, AMNH; Km 145, Oaxaca Hwy. 175, N of Oaxaca, 4000', CNC. *Puebla*: Tehuacán, PM, USNM. *Veracruz*: Atoyac, BMNH; N. Chocamán, USNM; Córdoba, BMNH, UCD, USNM; 14 mi E. Cuitláhuac, UCB; Jalapa, BMNH, OX, USNM; Misantla, PM; Orizaba, BMNH, LACM, PM, USNM; Paso San Juan, USNM; 16.2 mi N. Puente Nacional, UCB; Salto Eyipantla [8 km S. San Andrés Tuxtla], UCB. *Yucatán*: Chichen Itzá, CM; Valladolid, BMNH. **BELIZE**: *Toledo*: Punta Gorda, BMNH. **GUATEMALA**: *Alta Verapaz*: Chacoj, BMNH. *Guatemala*: Guatemala City, BMNH, LACM. *Huehuetenango*: 20 mi NW Huehuetenango, USNM. *Izabal*: Cayuga, USNM. *Retalhuleu*: San Felipe, Volcán, Santa María, 2000', BMNH. *Santa Rosa*: Barberena, BMNH. *Zacapa*: La Unión, 850 m, LACM. **HONDURAS**: *Atlántida*: Tela, CU. *Cortés*: La Cambre [=La Cumbre]. BMNH. *Distrito Central*: Tegucigalpa, USNM. **NICARAGUA**: "Chontales," OX. **COSTA RICA**: *Alajuela*: El Angel waterfall, 1350 m, 8.2 km downhill Vara Blanca, UPP; Finca Campana, 5 km NW Dos Ríos, 750 m, UPP. *Guanacaste*: 4 km E Casetilla, Rincon Nat. Pk., 750 m, UPP; La Mariksa Hda., Orosi, 550 m, UPP; Santa Rosa Nat Pk., 300 m, UPP. *Heredia*: La Selva Biol. Sta., 40 m, Puerto Viejo de Sarapiqui, UPP. *Cartago*: Juan Viñas, BMNH, CM, USNM; Orosi, 1200 m, PM; Sitio [de Avance ?], BMNH, CM; Tuis, CM. *Puntarenas*: Monteverde, LACM, UPP. *San José*: Candelaria Mts., BMNH. **PANAMA**: *Chiriquí*: Bugaba, BMNH; Lino, USNM; Potrerillos, 3600', USNM; Volcán de Chiriquí, 2-3000 ft BMNH. *Panamá*: Panama City, MCZ; Taboga Island, BMNH. **COLOMBIA**: *Antioquia*: Nari [Nare River], USNM. *Boyacá*: Muzo, 400-800 m, PM. *Cundinamarca*: Bogotá, BMNH; Finca San Pablo, 3 mi N. Albán, 1800 m, AMNH. *Magdalena*: Cacagualito, 1500', [hacienda 20 mi E. Sta.

Marta], BMNH; Don Amo, 2000', BMNH; Onaca, 220-2500', Sta. Marta, BMNH. **VENEZUELA**: *Aragua*: Carretera Maracay-Choroní, 1400-1575 m, UCV; La Isleta, Choroní, 200 m, UCV; La Victoria, 1700 m, UCV; Güiripa cerca San Casimiro, 780 m, UCV; Pozo del Diablo [Maracay], UCV; Rancho Grande, 1100 m, BMNH, UCD, UCV, USNM. *Carabobo*: San Estebán nr. Puerto Cabello, UCV. *Distrito Federal*: Caracas, CM; Cumbre de Boquerón frente a Bajo Seco, 1600 m, UCV.; La Guaira, ca. 1300', below Zigzag Station, OX. *Lara*: Terepaima, UCV. *Mérida*: La Mucuy, 2270 m, UCV; Mérida, BMNH, USNM. *Miranda*: Núcleo El Laurel, 1200-1300 m, UCV; Parque Nacional Guatopo, [El Lucero] 24 km N. Altagracia de Orituco, 640 m, UCV; Parque Nacional Guatopo, La Macanilla, 500 m, UCV. *Monagas*: Jusepín, UCV. *Yaracuy*: Aroa, USNM; Lagunita de Aroa, CM; Yumare, 50 m, UCV. **GUYANA**: East Demerara-W. Coast Berbice: MacKenzie, Demerara River, CU. **SURINAM**: *Marowijne*: Aroewarwa Kreek, Maroewym Valley, BMNH. *Suriname*: Geldersland, Surinam River, USNM. **FRENCH GUIANA**: *Guyane*: Cayenne, BMNH. **BRAZIL**: *Bahía*: Bahía [=Salvador], VM. *Mato Grosso*: Buriti dos Guimarães, 1200 m, CNC. *Pará*: Obidos, NMB; Cametá, BMNH; Pará [=Belém], BMNH; Santarém, USNM; Rio Tapajós, USNM. *Pernambuco*: Recife, BMNH. *Rio de Janeiro*: Nova Friburgo, PM.

Macrocneme chrysitis (Guérin-Méneville)

(Figs. 10-16, 136-137; Map 6)

Glaucopis chrysitis Guérin-Méneville, 1844:502
Euchromia (Macrocneme) chrysitis.- Walker, 1854:251
Macrocneme chrysitis.- Möschler, 1878:634.- Kirby, 1892:129
Macrocneme iole Druce (*nec* Druce, 1884), 1897:337 [misidentification]

This is a comparatively common species in Mexico and Guatemala, allied to *M. iole* Druce and *semiviridis* Druce, but geographically separated. The forewing iridescence is not as sharply delineated on the distal margin as in either of these species, but the hind tarsi are black, and the genitalic similarity suggests a close relationship among the three. The juxta is longer than in *iole* and its apical margin is spined on the left rather than on the right. The aedeagus has one large spine and lacks the spatulate process seen in *iole* and *semiviridis*. The females are unusual in having the ventral origin of the ductus seminalis originate at the base of the accessory bursa. A similar origin is found in females of *semiviridis*, while in *iole* and other species of *Macrocneme* the origin is dorsal.

MALE. *Head*: blackish brown, vertex and occiput metallic blue-green; labial palpi not reaching base of antennae. *Thorax*: blackish brown with metallic scaling coppery green; iridescence of disc clearly visible though somewhat obscured by overlying hairs on scutellum; metallic spots on patagia large, occupying most of sclerite between white points; tegulae with metallic streak on anterior mesal margin, white absent on underside; pectus blackish brown with metallic scales present only in coxal grooves, white spot absent on propleuron; legs blackish brown; iridescence as spots on forecoxae and as dorsal streaks on tibiae; hind tarsi entirely black. *Forewing*: dark blackish brown with basal half metallic coppery green, veins dark, apex slightly faded; oblique black streak from basal angle to bottom of cell; distal edge of metallic scaling sharply defined, touching end of cell; underside with blue-green to green metallic scaling in basal half, not reaching end of cell; inner margin gray-brown, retinaculum white. *Hindwing*: blackish brown with coppery iridescence restricted to patch below cell near Cu_{1+2}; underside metallic green except black-brown at apex, outer margin, and anal fold. *Abdomen*: shiny iridescent green, tergite I and mid-venter brownish black, scattered metallic scales on posterior margin of tergite I; pleura with three white spots diminishing in size caudad. *Genitalia*: as in Figs. 11-16 (drawn from RED prep. 39197, UCB, 9n); claspers of valvae somewhat asymmetrical, right side slightly longer and less curved than left side; inner surface of tips concave with inner edge of right process flatter and more elongate than corresponding edge on left process; mesal sclerite between clasper and ventral process of valva with prominent point; uncus skews left when viewed dorsally, base bilobed, margin at base of neck slightly protruding; juxta 5.5 mm long, apex rounded obliquely on left, incurved and quadrate on right, margin with 2-3 rows of uneven spines, tip reaching top of claspers; aedeagus with single large spine at left, raised margin at right, diaphragma sclerotized at attachment of aedeagus.

FEMALE. Essentially identical to male except for reduced metallic markings on patagia, tegulae, and venter of abdomen. *Genitalia*: as in Figs. 136-137 (drawn from RED prep. 39265, UCB, 5n); sternite VII triangulate, strongly sclerotized, center convex, caudal margin evenly curved with prominent notch at middle; intersegmental membrane VI-VII with paired sclerotized depressions from cephalic margin of sternite VII; inner fold of lamella antevaginalis strongly concave, margin at ostium evenly curved; sclerite of lamella postvaginalis spatulate and strongly decurved at ostium; ductus seminalis arising on ductus bursae ventrad to base of accessory bursa; stalk of accessory bursae thin, base plicate and folding left across ductus bursae; corpus bursae and signa as for genus.

VARIATION. Length of forewing: male, 18.3-20.8 mm; female, 16.8-20.9 mm. Membrane of corpus bursae occasionally without concentric plicae; notch on posterior margin of sternite VII varying from shallow to deep, but always present; hindwings occasionally with spur (R_s) from $S_c + R_1$ beyond middle; blackish brown ground color of wings may fade with age but brown hue always visible; pattern of wing iridescence constant, with only occasional differences such as female from Córdoba, Mexico, with forewing iridescence extending almost to tornus and basal black streak reduced to small mark below anal fold; color of iridescence on wings and thorax may vary between green and blue-green, but constant for individuals from one locality.

TYPE DATA. *chrysitis* Guérin-Méneville: Mexico, Bay of Campeche. Type specimen lost. According to P. Viette (*in litt.*) at the Paris Museum, Guérin's type material has been dispersed and there is nothing in Paris that is assuredly the type. The adequacy of Guérin's original description, including a definite type locality, leaves no doubt which species he described. Designation of a neotype is not required.

BIOLOGY. Powell and Chemsak (pers. comm.) experienced a massive influx of *chrysitis* to UV lights in September while collecting in Jacala (Hidalgo), Mexico. I have experienced the same phenomenon in March while collecting *iole* at San Vito, Costa Rica. This massing of a species at light has not been reported in other *Macrocneme*.

Females of *chrysitis* from Iturbide, Nuevo León, Mexico, were collected by J. A. Powell in 1976 and confined in polyethelene bags. Light green eggs, semispherical, shiny, smooth, and 0.92 mm wide, were deposited on the surface singly and in small clusters, and not in contact with each other. The chorion was transparent and possessed tiny hexagonal reticulations over the entire surface. First instar larvae emerged but failed to accept liliaceous leaves and codling-moth diet. Notable in one preserved specimen was the plumose condition of the primary setae.

GEOGRAPHICAL DISTRIBUTION. Eastern Mexico and south through Chiapas to Guatemala and Belize. Single records are available from Honduras and El Salvador.

FLIGHT PERIOD. Active throughout the year. Collection records are available for adults in every month of the year.

REMARKS. The British Museum has two males and a female labeled as collected by Jones from Castro, Paraná, Brazil. These undoubtedly have been mislabeled, and I have ignored the locality in this treatment.

Druce (1884) included in his description of *iole* a male from San Gerónimo, Guatemala, which I think should more accurately be considered a variant of *chrysitis*. The specimen has been mistakenly labeled as the type of *iole*. The genitalia (Figs. 10-12) does not correspond well to either *chrysitis* or *iole*. It appears to combine features of both species. The uncus dorsum and juxta (Figs. 10 and 11) resemble *iole*, while the single-spined aedeagus (Fig. 12) is similar to that in *chrysitis*. The specimen's appearance and distribution coincide most closely with *chrysitis*. Unless more material becomes available, I feel it should be considered a variant of *chrysitis* without formal taxonomic recognition.

I consider Druce's addition of Mexican localities to the range of *iole* to be in error and have included his reference (1897) in the synonymy of *chrysitis*, the species I believe was intended.

Costa Lima (1950), quoting Mabilde (1896), gives a larval description for "Macrocneme chrysitis." This must be a misidentification for the larva of *leucostigma* Perty, or one of the other lesser known species from Brazil since *M. chrysitis* does not occur in Brazil. I have treated the description and the host plants given by Mabilde as pertaining to *leucostigma* Perty.

Macrocneme affinis of Klages was treated as a subspecies of *chrysitis* by Draudt (1916) and Zerny (1931a), and as a synonym of *chrysitis* by Hampson (1914). I have examined the female holotype and find it bears no resemblance to *chrysitis*. It may be a synonym of *thyra*, or perhaps is related to *thyridia*, as suggested by Klages in his original description.

Macrocneme chrysitis never has white scales on the hind tarsi and is rarely encountered south of Guatemala. Hampson (1898) designated specimens from Guatemala and Rio Grande do Sul (Brazil) with white hind tarsi as "Ab. 1" of *chrysitis*. Unfortunately, Draudt (1916) named this "aberration", form *deceptans*. I have examined the Rio Grande specimen and find that it is a female of *leucostigma* Perty. I was unable to determine which specimen(s) Hampson had in hand from

Guatemala, but most likely the species was either *M. lades* or *cabimensis*. Both have white hind tarsi, and *lades*, especially, is commonly collected in Guatemala.

SPECIMENS EXAMINED (333): 138 males; 195 females. **UNITED STATES**: *Texas*, Bexar Co.: Ebony Hill Research Station, Mt. View Acres, 15 mi W. San Antonio, (Blanchard collection); Brewster Co.: Big Bend National Park, Basin area, CNC. **MEXICO**: *Chiapas*: El Bosque, CNC; Chiapas de Corzo, CNC; N. slope of Cerro Bola, N. Cerro Tres Picos, 1524-2134 m, CAS; Comitán, 10 mi NW., UCD; Cumbre de Arriaga, LACM; Huixtla, 20-25 mi N., CNC; Ocozocoautla, 2700', CAS; San Cristóbal de las Casas, 7200', CNC, LACM; Rizo de Oro, 6 mi SE., 2400', CAS; Simojovel, CNC; La Trinitaria, CNC. *Colima*: Volcan Colima, SMM. *Distrito Federal*: San Angel, AMNH. *Hidalgo*: Guerrero Mill, 9000 ft, BMNH; Chapulhuacan, 2 mi N., LACM; Jacala, 4500 ft, LACM, UCB, USNM; Jacala, 8 mi NE., 5200', LACM; El Salto, 15 mi SW. Jacala, 1800 m, UCB; Zimapán, 5 mi N., 2140-2280 m, CM. *Morelos*: "Morelos," PM; "Cautla" [=Cuautla], CAS; Cuernavaca, CAS. *Nuevo León*: Allende, 7 mi NW., UCB; 5 mi E. Galeana and 10 mi W. Galeana, UCB; Iturbide, 4 mi W., 5500', UCB; 18 and 20 mi W. Linares, UCB; Mesa de Chipinque, nr. Monterrey, 1365 m, CU; 4300', UCB; Monterrey, 5 mi S., CNC; Sabinas Hidalgo, 20 mi S., UCD. *Puebla*: Est. Puebla, Pumpstat III, 950 m, SMM; Necaxa, SMM; Puebla, AMNH; Villa Juárez, 1100 m, CNC, SMM. *Querétaro*: San Juan del Río, 7 mi E., 7300', CNC. *Sinaloa*: El Palmito, 8 mi W., UCB. *San Luis Potosí*: Ciudad de Maíz, 4 & 5 mi E., 4700', LACM; El Bonito, 7 mi S. Ciudad Valles, elev. 300', CAS; Ciudad Valles, 30 mi S., LACM; Tamazunchale, AMNH. *Tamaulipas*: Ciudad Victoria, CAS; Llera, CAS; Soto la Marina, 17 mi W., 400 m, CM. *Vera Cruz*: Coatepec, BMNH; Coscomatepec, LACM; Córdova, BMNH, UCD, USNM; Cuesta de Misantla, BMNH; Fortín, LACM; Fortín de las Flores, UCB; Huatuxco, BMNH; Jalapa, AMNH, BMNH, CM; Misantla, BMNH, PM, VM; Orizaba, AMNH, BMNH, CM, LACM, USNM, SMM; Poza Rica, 7 mi SW., 200', USNM; Puente Nacional, 6 mi SE. Rinconada, UCB; Teocebo, AMNH, CM; Vera Cruz, BMNH,PM. Not located: Choli (Sadler), BMNH. "Mexico," BMNH, CM, USNM; "States of Vera Cruz and Morelos" (Guérin), PM. **GUATEMALA**: *Baja Verapaz*: S. Gerónimo, BMNH. *Guatemala*: Ciudad de Guatemala, BMNH. *Jalapa*: Cayetano, PM. *Santa Rosa*: Barberena, 1300 m, BMNH. *Zacapa*: La Unión, 850 m, LACM. "Guatemala" (Guérin), PM. **BELIZE**: *Toledo*: Columbia, BMNH; Punta Gorda, BMNH; Río Grande, AMNH, BMNH; Río Temas [=Río Temash], BMNH. **HONDURAS**: "Honduras," USNM; Siguatepeque, 6 km N., UCB. **EL SALVADOR**: San Salvador, USNM.

Macrocneme iole Druce,   revised status

(Figs. 17-22, 138-139, 204-205; Map 7)

Macrocneme iole Druce, 1884:48, pl. VI, fig. 17.- [as synonym of *chrysitis*, Hampson, 1898:319; Draudt, 1916: 104; Zerny, 1931b: 241]
Macrocneme chrysitis.- Forbes (*nec* Guérin), 1939:129 [misidentification]

M. iole has been in synonymy with *chrysitis* Guérin since Hampson (1898). Its allopatric distribution and diagnostic genitalia convince me it is a valid species. It stands between *chrysitis* and *semiviridis* in a species group that extends from Mexico to Venezuela and Ecuador. It is probably more closely related to *semiviridis* by distribution and genitalic similarity than to *chrysitis*. All three share black hind tarsi and have a forewing iridescence that is sharply defined distally. The demarcation is often more pronounced in *iole* and *semiviridis* than in *chrysitis*. In the male genitalia, while *iole* has a small spine and spatulate process on the aedeagus, *chrysitis* has only a large spine, and *semiviridis* has only a pronounced spatulate process (cf. Figs. 12, 16, 20a-b, 26).

MALE. *Head*: dark brownish black, small lateral blue points on occiput; labial palpi not reaching base of antennae. *Thorax* (including pectus and legs): brownish black, iridescent markings blue; white absent from propleuron and from beneath tegulae; metallic points on anterior margin of tegulae, iridescence not extending along mesal margin; metallic scales of pectus restricted to femoral grooves; forecoxae mostly white in front with distal margins brownish black; hind tarsi brownish black. *Forewing*: dark brownish black with basal half golden-green; veins lightly black; diffuse, oblique black streak from basal angle to costa; distal edge of metallic scaling straight, well-defined, not touching end of cell; underside with basal half golden-green except where wings overlap; distal margin of metallic scaling less well-defined than on upper side. *Hindwing*: dark brownish black with golden-green spot in discal area; underside golden-green with apex, outer margin, and anal fold brownish black. *Abdomen*: dorsum (including tergite I) golden-green with faint, thin line of shiny scales mid-dorsally at base; underside brownish black with lateral margins metallic green. *Genitalia*: as in Figs. 17-22 (drawn from RED prep. 39181; USNM, 6n); dorsal processes (claspers) of valvae asymmetrical; inner surface of tips concave, with outer edge of right arm flatter and more elongate than corresponding edge on left arm; mesal sclerite between clasper and ventral process of valve with prominent point; uncus skews left when viewed dorsally, base bilobed; juxta 3.6-4.3 mm long, lateral margins narrowed beyond middle with apex twisting inward and not reaching top of claspers; tip margined by 2-3 rows of uneven spines; aedeagus with small spine on left and concave, spatulate process on right.

FEMALE. Essentially identical to male except occiput entirely blue rather than with two small points; forecoxae brownish black with white restricted to small proximal spots in front; metallic spot in discal area of hindwing reduced to scattered scales. *Genitalia*: as in Figs. 138-139 (drawn from RED prep. 244; LACM, 5n); sternite VII of sterigma triangular, caudal margin even, slight mesal emargination; intersegmental membrane between sternites VI and VII sclerotized but not as distinct pockets; sclerite of lamella antevaginalis trapezoidal with center convex and lower margin slightly decurved at ostium; dorsal wall of ductus bursae with thickened fold; membrane of bursa copulatrix not in concentric plicae; ductus seminalis arising dorsolaterad near base of accessory bursa.

VARIATION. Length of forewing: male, 16.5-18.5 mm; female, 17.0-20.0 mm. Vertex occasionally white scaled; white of forecoxae entire in males but reduced to small distal spots in females; amount of iridescence on the wings and body may vary within a population, e.g., among a long series of normal-appearing specimens taken at San Vito was one female (Fig. 205) with metallic scales absent from the head and thorax, wing iridescence reduced to a weak discal streak in the forewing above and to thin, basal blue streaks on both wings below, and a dark green abdomen devoid of iridescence except along lateral margins of tergites II and III.

TYPE DATA. ! *iole* Druce: Bugaba, Panamá, 800-1500 ft (Champion), male lectotype, **by present designation**, BMNH.

While Druce did not designate a type for *iole*, he did specify that the figure accompanying the original description was taken from the Chiriquí [Panama] specimens. The number of syntypes was not given, but five localities were listed: two in Guatemala (San Gerónimo and Calderas), one in Nicaragua (Chontales), and two in Panamá (Volcán de Chiriquí, 2000-4000 ft, and Bugaba).

When Hampson (1898) synonymized *iole* with *chrysitis* he followed Druce and listed 'type *iole*' as coming from Chiriquí. Apparently no specimen was labeled to reflect this designation. Since both Volcán de Chiriquí and Bugaba are in the province of Chiriquí, I have selected the male syntype labeled 'Bugaba, 800-1500 ft, Champion' as the *iole* lectotype.

BIOLOGY. Adults are seldom seen during the day but are attracted to UV light, often in huge numbers. (See *chrysitis* Biology for further comment.) The larva is unknown.

GEOGRAPHICAL DISTRIBUTION. Costa Rica and Panamá, with occasional records from Nicaragua and Colombia.

FLIGHT PERIOD. Probably active throughout the year, with collection records available for adults in every month except April and October.

REMARKS. The male syntype of *iole* from San Gerónimo, Guatemala, is problematical. The genitalia does not conform in detail to what I have illustrated for *iole* from Panamá, nor is it unmistakably like *chrysitis* from Mexico (compare Figs. 10-12 with 13-22). Since all other specimens of *Macrocneme* from Guatemala appear to be examples of *chrysitis*, I am treating the *iole* syntype as a unique variant of *chrysitis*. (See *chrysitis* Remarks for further comment).

SPECIMENS EXAMINED (190): 73 males; 117 females. **NICARAGUA:** *Chontales*: "Chontales," BMNH; Puente Quinama, E. of Villa Somoza, LACM. *Zelaya*: [El] Recreo, LACM. **COSTA RICA:** *Alajuela*: San Mateo, 1-2000 ft, USNM. *Cartago*: Juan Viñas, BMNH, CM; Orosí, 1200 m, BMNH; Tuis, BMNH, CM, USNM; Turrialba, USNM. *Guanacaste*: Cañas, 4 km NW. at Finca La Pacifica, UCB; Finca Jiménez Agricultural Station near Finca Taboga, LACM. *Heredia*: Carri Blanco [=Carriblanco], BMNH. *Puntarenas*: Avangarez [=Aranjuez], USNM; Monteverde, LACM; Parrita, UCD; Osa Peninsula, 1.8 mi W. of Rincón, LACM; San Vito at Finca Las Cruces, 3500 ft, UCB, UCD, USNM. *San José*: Candelaria Mts. BMNH; Carillo [=Carrillo], BMNH, CM; San José, BMNH; San Isidro del General, 14 km N., 1400 m, LACM; La Uruca [=Uruca], near San José, 1100 m, USNM. "Costa Rica" (Underwood, Cooper, Serre), BMNH, CM, PM. **PANAMA:** *Chiriquí*: "Chiriqui," BMNH, CAS, CU, USNM, VM; Bugaba, BMNH; El Volcán, AMNH; Lino, 800 m, PM, USNM; Volcán de Chiriquí, 2-3000 ft, BMNH; Potrerillos, 3600 ft, USNM; Rovira, SE. of Potrerillos Arriba, 1900 ft, UCB, USNM. *Canal Zone*: Barro Colorado, AMNH, CAS, MCZ; Madden Dam, MCZ. *Panamá*: Cerro Campana, UCB; La Victoria, MCZ. *Veraguas*: "Veragua," BMNH. "Panama," BMNH, CAS, CU. **COLOMBIA:** *Cundinamarca*: Bogotá, BMNH. *Valle*: Cartago, Caucatal, 600 m, PM.

Macrocneme semiviridis Druce,   revised status

(Figs. 23-26, 140-141, 206; Map 8)

Macrocneme semiviridis Druce, 1911:287.- [as synonym of *aurata* Walker] Forbes, 1939:128.

This species has the distal edge of the forewing iridescence sharply defined. The color of the iridescence varies geographically and tends to parallel similar variation in *M. oponiensis* and *cabimensis* where the three species are sympatric in Colombia and Ecuador. *M. semiviridis* has a South American distribution but is most closely allied by genitalia to the Central American and Mexican species *M. iole* and *chrysitis*. Males of *semiviridis* lack a spine on the aedeagus but have a spatulate process similar to that in *iole*. Sternite VII in the female sterigma is triangular and symmetrical as in *iole* and *chrysitis*. The unusual ventrolateral origin of the ductus seminalis is found elsewhere only in *chrysitis* among the described members of *Macrocneme*.

MALE. *Head*: brownish black; labial palpi reaching base of antennae. *Thorax*: brownish black; disc with iridescent blue scales obscured somewhat by overlying non-metallic scales, especially anteriorly; patagia with blue points adjoining lateral white spots; tegulae with white scales absent from underside, blue iridescence restricted to two small points on anterior margin; pectus without iridescence, white spot absent from propleuron; legs brownish black with metallic scales lining coxal grooves and as streaks on dorsal margins of tibiae; tips of hind tarsi black. *Forewing*: brownish black; basal half entirely metallic blue with distal edge of iridescence sharply delineated; underside similar, anal area in wing overlap light brown, non-iridescent; small accessory cell formed by R_2 anastomosing briefly with R_{3-5}; M_2 separate from M_3. *Hindwing*: basal and discal areas metallic blue, apex brownish black; costal margin in wing overlap non-metallic, light brown; underside more extensively iridescent than upperside, apex and outer margin brownish black. *Abdomen*: basal tergite brownish black, remaining tergites dark green, iridescence dull; venter with mid-ventral series of white spots and iridescent points only along lateral margins. *Genitalia*: as in Figs. 23-26 (drawn from RED prep. 39203; USNM, 3n); dorsal processes (claspers) of valvae mostly symmetrical, but inner margin of right arm wider and more emarginate than left arm; ventral processes recurved; dorsum of uncus evenly round, symmetrical (cf. *iole, chrysitis*); juxta an elongate process, narrowing beyond middle with right edge drawn out and curved; tip prominently spined; aedeagus unusual for absence of well-defined spine.

FEMALE. Similar to male except iridescence on underside of wings similar to upperside; metallic scaling of abdomen less pronounced, appearing diffuse or absent. *Genitalia*: as in Figs. 140-141 (drawn from RED prep. 39204; SMM, 2n); sternite VII of sterigma triangular, caudal margin slightly convex, emarginate at center; intersegmental membrane between VI and VII sclerotized with shallow pockets broadly separate laterally; sclerite of lamella postvaginalis shield-shaped, strongly convex medially, margin decurved and inserted at left of ostial opening; caudal margin sternite VIII deeply emarginate; blind pouch from right of ductus bursae; stalk of accessory bursa extending left across ductus bursae; ductus seminalis arising ventrolaterad below stalk of accessory bursa.

VARIATION. Length of forewing: males, 16.-19.8 mm, females, 16.5-19.7 mm. White irrorations occasionally on the vertex; labial palpi not always reaching antennal bases; iridescent points on anterior margins of patagia and tegulae occasionally coalescing;M_2 of forewing may be connate with M_3; M_3 of hindwing may be short-stalked with Cu_{1+2}; iridescence of wings varies from steel blue to blue-green to yellow-green, with color usually constant on individual specimens and within a population; iridescence in northern Venezuelan populations steel blue, in central Colombia almost blue-green, and in Ecuador yellow-green; occurrence of black scales at base of forewing more often in females than males; tip of juxta broadly bifurcate on occasion; caudal margin of sternite VII in female genitalia slightly emarginate mesally, as in Fig. 140, or strongly notched.

TYPE DATA. ! *semiviridis* Druce:female holotype, Colombia [Risaralda], Siató, Río Siató; slopes of Chocó, 5200 feet, BMNH.

BIOLOGY. Unknown.

GEOGRAPHICAL DISTRIBUTION. Colombia, Ecuador, and north-central and western Venezuela.

FLIGHT PERIOD. Probably active throughout the year, with collection records for every month except November.

REMARKS. Forbes (1939) misidentified this species and incorrectly synonymized it under *aurata* Walker which he transferred to *Macrocneme* from *Calonotos*. His illustration and description of "aurata" are of a male of *M. cabimensis* Dyar. *M. semiviridis* is not recorded from Panamá, nor are its hind tarsi ever white-tipped. The tympanic organ of *semiviridis* is illustrated in Kiriakoff (1948).

SPECIMENS EXAMINED (62): 36 males, 26 females. **COLOMBIA**: *Antioquia*: Nari [=Nare] River, USNM; Medellín, La Estrella, 1700 m, SMM. *Boyacá*: Muzo, 400-800 m, PM. *Cundinamarca*: Finca San Pablo, 3 km N. Albán, 1800 m, AMNH; Cananche, BMNH; Chocó, near Bogotá, USNM; Guaicarmo, USNM; Monterredondo, 1420 m, SMM. *Meta*: Río Negro nr. Villavicencio, USNM; Río Negro, E. Colombia, 800 m, BMNH, PM. *Risaralda*: Distrito de Pereira, BMNH; [holotype] Río Siató, 5200 ft, BMNH. *Santander*: Alto Río Opón, 900-1100 m, SMM; Landázuri, 800-850 m, AMNH. *Tolima*: Honda, BMNH; Labora, Quindío Pass, 2700 m, PM. *Valle*: Río Dagua, BMNH; Funtas, Río Dagua, W. Colombia, 400 m, BMNH. **ECUADOR**: *Chimborazo*: Chimbo, BMNH. *El Oro*: Zaruma, BMNH, USNM. *Guayas*: Bucay, HM. *Pichincha*: Santo Domingo de los Colorados, USNM. **VENEZUELA**: *Aragua*: Rancho Grande, 1100 m, RML, UCD, UCV, USNM; carretera Maracay-Choroní, E. Aragua, 1100 m, UCV. *Barinas*: La Chimenea [nr. La Soledad], 1500 m, UCV. *Carabobo*: San Esteban, BMNH. *Miranda*: Parque Nacional Guatopo, Agua Blanca, 500 m, UCV.

Macrocneme cabimensis Dyar, new status

(Figs. 27-30, 142-143, 209, 234; Maps 9 and 10)

Macrocneme lades cabimensis Dyar, 1914:162.- [as synonym of *M. thyridia*] Forbes,
1939:130.- Hampson, 1914:387.
Macrocneme aurata.- Forbes (*nec* Walker, 1854), 1939:128 [misidentification]
Macrocneme leucostigma.- Druce (in part, *nec* Perty, 1833), 1884:47.- Druce, 1896:336
[misidentification]

The polytypic nature of the wing iridescence in *cabimensis* is largely responsible for
its misidentification. The male genitalia, especially the deeply concave tips on the
claspers and the strongly white hind tarsi, are diagnostic. In Central America the wing
iridescence is notably reduced, resulting in an appearance similar to *lades* Cramer. In
Colombia and Ecuador *cabimensis* is involved in a mimetic complex with *M. semiviridis*
Druce and *oponiensis* Dietz. Here the three species can only reliably be separated by the
male genitalia or by differences in the hind tarsi, which are broadly white in *cabimensis*,
black in *semiviridis*, and have a thin white streak on the inside margin of the distal
segments in *oponiensis*. Specimens from Belize to Panamá conform to the following
descriptions (see Fig. 234). Individuals from Colombia and Ecuador vary in having the
basal half of the wings covered by iridescent scales and the color may also differ (Fig.
209).
MALE. *Head*: dark brownish black; labial palpi not reaching base of antennae,
scattered white scales or tiny streak laterally on segment II. *Thorax*: brownish black, disc
with metallic spots obscured by overlying non-metallic hairs; iridescent points of patagia
adjoining lateral white points; tegulae with small, metallic points on anterior margin,
white scales underneath not visible along mesal fringe; pectus with brownish black hairs
interspersed with white spots on pro- and metapleuron; legs brownish black; coxae,
trochanters, and femoral grooves white; metallic scales lining femoral grooves and
forming streaks on outer surface of tibiae; hind tarsi strongly white-tipped. *Forewing*:
dark brownish black with metallic blue spots at base, oblique black streak from basal
angle to bottom of cell, bordered distally by transverse fascia of short blue streaks in
discal area; underside with basal half metallic blue, except tan and white along inner
margin where wings overlap; retinaculum white. *Hindwing*: brownish black with metallic
scales absent, costa gray-brown; underside mostly metallic blue-green except brownish
black at apex and anal fold. *Abdomen*: dorsum (except tergite I) and pleura iridescent
green, venter mostly white, caudal margins on sternites IV-VIII brownish black laterally.
Genitalia: as in Figs. 27-30 (drawn from RED prep. 39219; USNM, 2n); valvae
symmetrical; tips of dorsal processes distinctively concave (Fig. 27a), ventral edge slightly
broader and more elongate than dorsal edge; dorsum of uncus symmetrical, slightly
flattened with two small unequal flanges along dorsolateral margins; juxta elongate, with
left side drawn out into long process that reaches top of claspers; tip prominently spined;
aedeagus with single small spine at right and faint carina on left above point of
attachment to diaphragma.

FEMALE. Similar to male except: *Thorax*: white scarce on underside of tegulae; metallic scales on anterior and mesal margin of tegulae continuous rather than two distinct points; white on pectus restricted to spots on coxae and trochanters and absent from femora. *Abdomen*: iridescence blue-green, white on venter in mid-ventral series of spots, large basally and diminishing in size distally. *Genitalia*: as in Figs. 142-143 (drawn from RED prep. 39242, USNM, 2n); sternite VII as inverted triangle; intersegmental membrane between sternites VI and VII sclerotized, forming single crescent-shaped pouch; sclerite of lamella postvaginalis trapezoidal; blind pouch absent from wall of ductus bursae; ductus seminalis arising dorsocephalad on base of stalk for accessory bursa.

VARIATION. Length of forewing: males, 16.8-19.6 mm; females, 18.4-19.4 mm. Labial palpi occasionally reaching base of antennae; white scales sometimes absent from segment II of labial palpi and underside of tegulae, especially in females; color and amount of wing iridescence geographically variable (see Remarks); black streak and medial fascia of forewing variable in width; hindwings, especially in non-Panamanian specimens, with scattered metallic scales in basal and discal areas.

TYPE DATA. ! *cabimensis* Dyar: Panamá, Cabima, May 1911 (Busck), two female syntypes, USNM. I have labeled one as the lectotype.

BIOLOGY. An adult male has been collected visiting *Heliotropium indicum* in Panamá. Drying specimens of this plant are known to attract numerous species of ctenuchids (Hagmann, 1938), but *Macrocneme* species seldom are encountered.

GEOGRAPHICAL DISTRIBUTION. The Gulf side of Central America from Belize and Guatemala to Costa Rica, Canal Zone, western Panamá, central and western Colombia, and north-central Ecuador.

FLIGHT PERIOD. Probably flies throughout the year. Collection records are available for specimens in every month of the year except October.

REMARKS. Forbes apparently never examined Walker's type of *aurata*. He transferred the name without comment from *Calonotos* to *Macrocneme* and then described and illustrated the male genitalia of *M. cabimensis*.

The specimens listed by Druce (1884, 1896) as *M. leucostigma* from Río Sarstoon [=R. Sarstún], British Honduras, and Emperador, Panamá, are females of *M. cabimensis*.

Possibly *M. cabimensis* forms a mimetic complex with *M. oponiensis* and *semiviridis*. When the three species are sympatric in Colombia and Ecuador, the pattern and color of the wing iridescence are almost identical. The basal half of the wings is entirely metallic and is seldom interrupted by the usual thin streak of black scales from the basal angle. The hindwing is also largely iridescent except at the apex. The color of the iridescence varies geographically (blue to blue-green in Colombia; bright yellow-green in Ecuador), but each species changes accordingly so that a similarity in appearance is maintained. Where *M. cabimensis* occurs out of the range of *semiviridis* and *oponiensis* in Central America, the wing iridescent pattern is reduced to short basal streaks and a narrow medial band on the forewing and almost no metallic scaling on the hindwing. The appearance becomes more like *lades* Cramer. Dyar reflected this observation by naming *cabimensis* originally a subspecies of *lades*. Whether there is any biological significance to the change in *cabimensis* when it is no longer sympatric with *oponiensis* and *semiviridis* is unknown. Perhaps the appearance is related to the occurrence of appropriate models.

SPECIMENS EXAMINED (95): 41 males; 54 females. **BELIZE**: *Toledo*: Punta Gorda, BMNH; Río Grande, CM; Río Sarstoon [=R. Sarstún], BMNH. "British Honduras," BMNH. **GUATEMALA**: *Izabal*: Barrios [=Puerto Barrios], sea level, BMNH; Cayuga, BMNH, CM; Quiriguá, BMNH. **HONDURAS**: *Cortés*: La Cambre [=La Cumbre], BMNH. **COSTA RICA**: *Alajuela*: Finca La Campana, El Ensayo, 7 km NW Dos Ríos, 700 m, UPP. *Guanacaste*: Santa Rosa Nat. Pk., 300 m, UPP. *Heredia*: La Selva Biol. Sta., Puerto Viejo de Sarapiqui, 40 m, UPP. *Limón*: Cerro Tortuguero, N. edge Tort. Nat. Pk., 0-100 m, UPP; Guápiles, USNM; La Lola [Experiment Station, 50 m], USNM. **PANAMA**: Isthmus of Panamá, BMNH. *Canal Zone*: Cocoli, CM; Emperador, BMNH; Gatún, BMNH; USNM; Piña area, MCZ. *Colón*: Colón, Santa Rita, 1500', USNM. *Panamá*: Araihan [Arraiján], BMNH; Cabima, USNM; La Victoria, Cerro Azul, MCZ; Cerro Campana, 3000', MCZ, USNM; Punta Vique [=Bique], BMNH. **COLOMBIA**: *Antioquia*: Nari [=Nare] River, USNM. *Boyacá*: Muzo, 400-800 m, VM; Muzo, Río Cantinero, 400 m, BMNH. *Cundinamarca*: Bogotá, BMNH, VM; Cananche, BMNH. *Risaralda*: Santa Cecilia, 800 m, CM. *Valle*: Río Dagua, BMNH. **ECUADOR**: *Imbabura*: Paramba, Río Mira, 3500 ft, *in copulo* pair, OX; Paramba, 3500-4000', BMNH, USNM. *Pichincha*: Cayambe, 9000 ft, *in copulo* pair, USNM.

Macrocneme oponiensis Dietz, new species

(Figs. 31-34, 144-145, 207; Map 10)

The genitalia of this species are diagnostic, and the adult can be identified readily by the restriction of white to a thin band on the inner margin of the hind tarsi (see Fig. 207). It belongs to a species group including *semiviridis* and *cabimensis* that has the distal edge of the forewing iridescence sharply delineated. The color of the iridescence varies geographically and follows the same patterns found in *semiviridis* and *cabimensis*, i.e., it is blue to blue-green in Colombia and bright yellow-green in Ecuador.

MALE. Length of forewing: 18.5 mm. *Head*: brownish black to black, occiput with two metallic spots; labial palpi not reaching base of antennae. *Thorax*: brownish black, disc metallic blue with scales obscured somewhat by overlying non-metallic hairs; patagia with blue spots adjoining lateral white spots; tegulae with blue streak across foremargin, white absent on underside; pectus without iridescence; white spot on propleuron reduced to a few scales, appearing absent; legs brownish black, forecoxae with proximal white spot large and metallic scales absent, coxal grooves lined with metallic scales, fore and mid tibiae with thin metallic streaks on outer margin, tips of hind tarsi streaked with white on inner margin. *Forewing*: black to brownish black, basal half iridescent blue with distal edge sharply delineated, black scales absent at base; underside similar to upperside. *Hindwing*: basal and discal areas iridescent blue, apex and outer margin black to brownish black; underside similar. *Abdomen*: dorsum dark green with tergites II and III iridescent blue-green, venter brownish black with mesal white spots large on sternites II and III, small on sternites IV and V, lateral white spots on basal sternite, single spot in pleura; bilobed pocket from anterior margin of sternite VIII. *Genitalia*: as in Figs. 31-34 (drawn from paratype, RED prep. 39227, SMM, 2n); dorsal processes (claspers) of valvae somewhat asymmetrical, arms clavate toward tips, dorsal margins as in Fig. 31a; mesal sclerite without knob; dorsum of uncus strongly skewed left when viewed from above, lateral margins asymmetrically flanged, left side broader and more horizontal than right side; juxta extends well above base of ventral processes, right margin incised, tip spinose; dorsum of aedeagus with two spines, left small, straight, right large, S-shaped; vesica of three membranous bursae, diaphragma sclerotized at attachment of aedeagus.

FEMALE. Essentially identical to male, including diagnostic white streak on inner margin of hind tarsi. *Genitalia*: as in Figs. 144-145 (drawn from allotype, RED prep. 280, USNM, 2n); sternite VII of sterigma u-shaped, apical margin skewed left, not continuous with sclerotized intersegmental membrane or large invaginated pocket at right; inner fold of lamella antevaginalis as wide as overlying fold (sternite VII); sclerite of lamella postvaginalis bent medially, only slightly inserted at ostium; dorsal wall of ductus bursae unmodified, membrane plicate.

VARIATION. Length of forewing: males, 16.5-19.0 mm; females, 17.0-20.0 mm. Color of wing and body iridescence varies widely depending on population. Blue or bluish green scales are typical of the Colombian specimens, while in the Ecuadorian populations the scales are yellow-green. Metallic color of thorax and abdomen varies in accordance with wing iridescence. Presence or absence of iridescence on forecoxae and hind tibiae varies within populations.

TYPE DATA. Male holotype: Alto [upper] Río Opón, Santander/Sur, Colombia, 900 m, 8-II-49 (Richter), SMM. Allotype: Villavicencio, [Meta], Colombia, VI-1919, (Apollinaire), Dognin Collection, USNM. Paratypes (24):   15 males, 9 females. COLOMBIA: *Santander/Sur*: La Carmen, 900 m, March (Richter), SMM; Río Carare, 1000 m, June (Richter), SMM; Alto Río Opón, 1000 m, February (Richter), SMM. ECUADOR: *Bolívar*: Balzapamba, October-February (deMathan), BMNH. *Esmeraldas*: Salidero, 350 ft, February (Fl. & Mik.), BMNH; San Mateo, August (Förster), SMM. *Guayas*: Bulim [=Berlim], 160 ft, January (Fl. & Mik.), BMNH. *Imbabura*: Paramba [Hacienda], 3500', dry season, March-May (Rosenberg), BMNH. *Pichincha*: Santo Domingo de los Colorados, 650 m, October (Förster), SMM.

BIOLOGY. No information.

GEOGRAPHICAL DISTRIBUTION. North-central Colombia and western Ecuador.

FLIGHT PERIOD. Adult collection records are available for all months except July and September, probably indicating that the species is available throughout the year like most *Macrocneme*.

ETYMOLOGY. The specific epithet is taken from the type locality.

Macrocneme immanis Hampson

(Figs. 35-38, 169-170, 210; Map 12)

Macrocneme immanis Hampson, 1898: 320-321, pl. XII, fig. 1

A comparatively large *Macrocneme* with a characteristic dull iridescent sheen of the wings. This is the most commonly collected member of a species complex that includes two new sympatric species, *M. habroceladon* and *tarsispecca*. The appearance and distribution of the three are so similar that only the male genitalia will separate them. *M. immanis* males differ slightly externally by having more extensively white hind tarsi and prominent white scaling on the underside, especially basad on the abdominal venter. The identification of females is difficult due to the lack of characters for properly associating the sexes within the complex. The structure of sternite VII is sufficiently different to distinguish *tarsispecca* from *immanis*. The female of *habroceladon* is unknown.

MALE. *Head*: blackish brown, occiput with iridescent green spots; labial palpi reaching base of antenna, segment II slightly porrect. *Thorax*: blackish brown, iridescence dull green; metallic scales on disc obscured by overlying hair-like scales; metallic spots on patagia small; tegulae with scattered metallic scales along anterior and mesal margins, white absent from underside; pectus brown interspersed with white, white spots on pleura above coxae and on metepimeron; legs blackish brown, inner margins of coxae, femora, and tibiae streaked with white, hind coxae and fringe of entire tarsi prominently white. *Forewing*: brown, basal half suffused with dark metallic green to end of cell, distal margin of iridescence not sharply delineated, basal black scales absent or scarce; underside similar, except iridescence greenish blue, depending on angle of light, tip of retinaculum white. *Hindwing*: brown with dark green iricescent patch in discal area; underside entirely suffused with blue-green iridescence, except narrowly at apex. *Abdomen*: dorsum suffused with dark green iridescence, shiny longitudinal striae absent; basal two segments of venter mostly white, remaining segments with small mesal spot, pleura with two white spots basally. *Genitalia*: as in Figs. 35-38 (drawn from RED prep. 39223, SMM, 5n); dorsal arms (claspers) of valvae symmetrical, inner margins without indentations; uncus skewed slightly to left when viewed above, dorsum round, lateral margins horizontally flanged; apex of juxta quadrate, extending slightly beyond base of ventral processes, spines absent; aedeagus with two spines, longest on left with surface prominently spinose, short stubby spine on right with apex occasionally blunt.

FEMALE. Essentially identical to male except: *Thorax*: less white on underside of pectus and legs; inner margins of coxae, femora, and tibiae dark brown with scattering of white scales along mesal margin of forecoxae and in hindcoxal grooves. *Abdomen*: white on venter restricted to mesal series of small spots, with larger pair laterally on basal segment; pleura white spotted on basal two segments, smaller spots on remaining distal segments. *Genitalia*: as in Figs. 169-170 (drawn from RED prep. 268, SMM, 2n); sternite VII of sterigma shield-shaped, apex cephalad, broad, skewed to left; caudal margin slightly convex mesad, with small independent pocket in intersegmental membrane; sclerotized band from margin of encircling pleurite at right; fold of lamella antevaginalis half as wide as overlying sternite VII; sub-ovoid sclerite of lamella postvaginalis decurved slightly along ostial opening; sternite VII with two prominent plicae, one unattached at left, the other appended from mesal prominence; ductus bursae and stalk of accessory bursa with concentric plicae; thickened plicae in dorsal wall of ductus bursae; opposing, scallop-shaped signa with recumbent spines in bursa copulatrix.

VARIATION. Length of forewing: males, 21.7-23.7 mm; females, 21.5-23.2 mm. Wings are normally brown suffused with a muted green iridescence, but occasionally the ground color is a brownish black (Río Pampas, Perú) or the iridescence is a muted blue (San Pedro, Argentina; Río Pampas, Perú). In males, the white scaling on the underside is variable, occasionally sparse or entirely lacking along the inner margins of the femora and tibiae of the forelegs, tibiae of the hindlegs, or as thin, transverse bands on abdominal venter (Chaco, Bolivia; Yungas de Incachaca, Bolivia).

TYPE DATA. ! *immanis* Hampson: male lectotype, **here designated**, Chaco, Bolivia (Garlepp), BMNH. One female with same data was included by Hampson in the original description.

BIOLOGY. No information.

GEOGRAPHICAL DISTRIBUTION. Central Perú to northern Argentina and east to the lowlands [=yungas] of Bolivia.

FLIGHT PERIOD. Probably active throughout the year. Collection records indicate that adults are found in every month except July and August.

REMARKS. The muted, dull appearance of the wing iridescence and the absence or scarcity of black scales on the forewing are good indicators whether *immanis, habroceladon*, or *tarsispecca* is present. Initially it is easier to separate *M. immanis* from its siblings by the whiter basal leg segments, the white abdominal base in males, and the white fringe of the hind tarsi extending to the tibia. Although the three species are often sympatric, distribution records show the range of *immanis* to extend only to the Department of Junín, Perú, while both *habroceladon* and *tarsispecca* occur further north in Colombia (Río Negro). Since the three occur together in localities as far apart as La Oroya, Perú, and the Yungas del Palmar, Bolivia, it is best to use genitalia characters in making an identification.

SPECIMENS EXAMINED (64): 46 males, 18 females. PERU: *Ayacucho*: Río Pampas, CAS. *Cuzco*: Cajón [Hacienda], BMNH; Marcapata, PM; Vilcanota, 3000 m, BMNH. *Junín*: Perené, CU. *Puno*: La Oroya [=Oroya], R. Inambari, 3100 ft, dry season, BMNH; Oconeque, Carabaya, 7000 ft, BMNH; S. Domingo, Carabaya, 6000-6500 ft, dry/wet season, BMNH; USNM. BOLIVIA: *Cochabamba*: Chapare, 400 m, LACM, SMM; Yungas de Incachaca, 2100 m, SMM; Yungas del Palmar, 2000 m, SMM; Cochabamba, CM. *La Paz*: Chaco, BMNH, USNM; Coroico, 1400, 1500, and 1900 m, BMNH, SMM, VM; Yungas, Chulumani, 1200 m, SMM; Yungas de La Paz, 1000 m, USNM; La Paz, 1000 m, BMNH; R. Solocame [=R. Solacama], 1200 m, BMNH; R. Tanampaya [=R.Tamampaya], BMNH; R. Songo [=R. Zongo], 750 m, PM. "Bolivia" (Schaus Collection), USNM. ARGENTINA: *Jujuy*: Jujuy, 1250 m, SMM; Ledesma, SMM; S[an] Juancito, CU; San Pedro, 580 m, SMM; Yala, 1450 m, SMM; Yuto, 350 m, SMM. *Salta*: Aguaray, IML; Dept. San Martín, Pocitos, IML;

Salta, 3300 ft, BMNH. *Tucumán*: Aconquija [sierra], 500 m, IML; El Cuartiadero, IML; Quebrada de Lules, IML; Tucumán, BMNH, PM, USNM; Villa Nougues, IML.

Macrocneme tarsispecca Dietz, new species

(Figs. 39-42, 133, 166-168, 212; Map 11)

A member of the species complex including *immanis* and *habroceladon* that is broadly sympatric from Colombia to Argentina, and unique in the genus for the tiny invaginated pocket of unknown function in abdominal sternite VI in both sexes (Figs. 133 and 168). *M. tarsispecca* is phenotypically indistinguishable from *habroceladon*, but is more commonly encountered and has upturned arms on the claspers and an incised juxta to separate it morphologically.

MALE. Length of forewing: 21.7 mm. *Head*: brownish black, occiput and vertex interspersed with metallic blue-green; labial palpi not reaching base of antennae. *Thorax* (including legs): brownish black; iridescence of disc blue-green, obscured by overlying hair-like scales; dorsal and lateral white spots of patagia small, intervening metallic spots reduced to few scales; metallic markings of tegulae reduced to few scales on anterolateral margin, white scales on underside not visible along mesal fringe; white scales on pectus limited to small spots at proximal ends of coxae and on propleuron and metepimeron; outer surface of fore and hind coxae white-spotted; coxal grooves and tibiae streaked with metallic blue-green; hind tarsi white-tipped. *Forewing*: brown, suffused to beyond middle with dull green iridescence, base flecked with blue; basal black streak below A_1 and at costa small, irregular; underside similar, but iridescence slightly bluer depending on light, extending slightly beyond cell, black scales absent. *Hindwing*: brown suffused with dull green iridescence except at apex and outer margin; underside entirely suffused with dull iridescent blue or blue-green scales, depending on angle of light. *Abdomen*: entirely iridescent dark green, not shiny; thin longitudinal bands not evident; white spots of basal tergite small; venter similar, except white of mesal series forming thin bands along anterior margin of segment; sternite VI with small invaginated pocket on left (Fig. 133); sternite VIII with membranous folds (coremata?) invaginated from anterior margin. *Genitalia*: as in Figs. 39-42 (drawn from RED prep. 39225, SMM, 2n); dorsal arms (claspers) of valvae symmetrical but distinctly curved upward beyond middle; uncus skewed strongly to left when viewed from above, lateral margins extended as asymmetrical flanges, broad on left, narrow and vertical on right; juxta long, extending well above base of ventral processes, apex prominently spinose at margin and sharply incised on right; dorsum of aedeagus with two small spines pointing in opposite directions.

FEMALE. Length of forewing: 21 mm. Like male except: *Thorax*: iridescence of disc green with occasional bluish sheen; white spots of forecoxae smaller, larger on hind coxae. *Forewing*: iridescence green to blue-green, basal black scales much reduced, basal blue flecks absent. *Hindwing*: iridescence absent on underside from apex and inner margin. *Genitalia*: as in Figs. 166-167 (drawn from RED prep. 269,

SMM, 1n); sternite VII of sterigma broadly shield-shaped, apex cephalad, broad; small independent ovoid sclerite intersegmentally between VI and VII; tiny flap and invaginated pocket (Fig. 168) on caudal margin of sternite VI; plicae associated with medial protuberance on segment VIII much reduced; fold of lamella antevaginalis half as wide as overlying sternite VII; irregular, ovoid sclerite of lamella postvaginalis not strongly decurved at ostium; ductus bursae and stalk of accessory bursa with concentric plicae; stalk of accessory bursa comparatively short; signa as for genus.

VARIATION. Length of forewing: males, 20.0-22.0 mm; females, 21 mm.

TYPE DATA. Holotype male: Bolivia, [*La Paz*], Sarampiuni, San Carlos, 1000 m, 31-8-50 (W. Förster, leg.), SMM. Allotype: Bolivia, [*Cochabamba*], Chapare-Gebeit [=upper region] Río Chipiriri, 400 m, 31-X-1953 (W. Förster, leg.), SMM. Paratypes (20): 19 males, 1 female(?). COLOMBIA: *Meta*: Río Negro, E. Colombia, 800 m (Fassl), BMNH. PERU: *Puno*: La Oroya[=Oroya], R. Inambari, wet season, 3100 ft, March, September (Ockenden), BMNH; Santo Domingo, Carbaya, wet season, 6000-6500 ft, November (Ockenden), BMNH; Tinguri [=Hacienda Tincuri], Carabaya, wet season, 3400 ft, January (Ockenden), BMNH. BOLIVIA: *Cochabamba*: Río Cristal Mayo [=R. Cristalmayo], Chapare, female, 1000-2000 m, October (Peña), CU; Yungas del Palmar, 1-2000 m, January, September, November (Zischka), SMM. *La Paz*: Prov. Inquisivi, Sacambaya, 6500 ft, May-October (Kettlewell), BMNH; Río Songo [=R. Zongo], 750-800 m (Fassl), PM, VM; Yungas, Chulumani, 1200 m (Schulze), SMM.

BIOLOGY. No information.

GEOGRAPHICAL DISTRIBUTION. Eastern Colombia, southeastern Perú, and lowlands of Bolivia.

FLIGHT PERIOD. Collection records show adults flying in January and March and again from August through November. Additional collecting will probably show that the species flies throughout the year, like other *Macrocneme*.

ETYMOLOGY. The specific epithet is taken from the Greek, m., *tarso*s =ankle; and the Anglo-Saxon *specca* =spot and refers to the hind tarsi.

REMARKS. Typically the females are difficult to associate, but I am assuming the allotype is properly placed here due to the small invaginated pocket on the right side of sternite VI (Fig. 168), seeming to correspond to a similar invagination on the left side of the same sternite (VI) in the male holotype (Fig. 133). The function of the pockets is unknown, but they may serve as receptacles for scent scales during mating. They are not found in males of *immanis* or *habroceladon*.

Macrocneme habroceladon Dietz, new species

(Figs. 43-46, 211; Map 11)

This is the least commonly collected of the three species making up the *immanis* complex. *M. habroceladon* occurs sympatrically with *tarsispecca* from Colombia to Bolivia, and with *immanis* from Perú to Bolivia. Its presence is best detected by an examination of the male genitalia. The small aedeagal spines, the large, non-incised juxta, and flattened scales interspersed among setae on the ventral processes of the valvae are diagnostic. Also, the abdominal venter is less white when compared to *immanis*.

MALE. Length of forewing: 21.5 mm. *Head*: blackish brown with small metallic spots on occiput; labial palpi reaching base of antennae. *Thorax* (including legs): blackish brown, metallic scales of disc obscured by overlying hair-like scales; patagia with mesal and lateral white spots small, lateral metallic spots reduced to few scales; tegulae without metallic scales, white absent from underside; pectus with white spot absent from propleuron but present on metepimeron; inner margin of forecoxae white, all coxae white-spotted proximally, spot smaller in hind legs; hind tarsi white-tipped. *Forewing*: brown, basal half to end of cell suffused dull green, iridescent; few black scales at basal angle and along costa; underside similar except iridescence not reaching end of cell; retinaculum brown. *Hindwing*: brown with dull green iridescent patch in discal area; underside mostly iridescent green, brown at apex. *Abdomen*: dull iridescent green with thin, shiny transverse line dorsomesad; venter blackish brown with lateral margins shiny green, basal segment with three white spots, remaining segments with single mesal spot; sternite VIII with anterior margin as in Fig. 132, possibly coremata? *Genitalia*: as in Figs. 43-46 (drawn from RED prep. 39224, SMM, 2n); dorsal arms (claspers) of valvae asymmetrical, left arm longer and narrower than arm on right; ventral processes with flat, bifurcated scales interspersed among setae at tip; uncus skewed left when viewed dorsally, dorsum round, lateral margins extended as asymmetrical horizontal flanges; juxta long, extending well above base of ventral processes of valvae, apex margined with small spines, short, thumb-like flap from right; dorsum of aedeagus with two small dextrad spines, left tip acuminate, right tip mucronate.

FEMALE. Not known.

VARIATION. Length of forewing: males (5n), 20.8-21.5 mm.

TYPE DATA. Holotype male: Cochabamba, Bolivia (J. Steinbach), Acc.[#] 6873; USNM Type #73264, USNM. Paratypes (16): all males. COLOMBIA: *Meta*: Río Negro, E. Colombia, 800 m (Fassl), BMNH. PERU: *Puno*: La Oroya [=Oroya], R. Inambari, September, 3100 ft, dry season (Ockenden), BMNH. BOLIVIA: *Cochabamba*: "Cochabamba" (Steinbach), CM, USNM; Yungas del Palmar, 2000 m, November (Zischka), SMM.

I have tentatively placed one female from Río Negro, Colombia (Fassl), BMNH, and one from Prov[encia] del Sara, Bolivia, 450 m (Steinbach), CM, under *habroceladon*. I am not certain that these are properly associated sexually and therefore do not consider them paratypes.

BIOLOGY. No information.

GEOGRAPHICAL DISTRIBUTION. Eastern Colombia, southeastern Perú, and the lowlands of Bolivia.

FLIGHT PERIOD. The activity of adults on a year-round basis is unknown, due to a scarcity of specimens. Adults have been collected in July, September, and Nov.

ETYMOLOGY. The specific epithet is taken from the Greek *habros* =pretty, and the French *celadon* =sea green, to reflect a distinguishing feature of the wing iridescence.

REMARKS. I know of no way to distinguish the females.

Macrocneme lades (Cramer)

(Figs. 47-53, 148-149, 196, 213-214; Maps 13 and 14)

Sphinx lades Cramer, [1776]:153 (Index); [1775]:131, pl. 83, fig. E
Zygaena ladis [*sic*].- Fabricius, 1781:165 [*lapsus*]
Macrocneme lades.- Hübner, 1818:15.- Kirby, 1892:129
Euchromia (Macrocneme) lades.- Walker, 1854:251
Euchromia (Macrocneme) aurata Walker, 1854:250 [NEW SYNONYMY]
Calonotos aurata.- Butler, 1876:369.- Hampson, 1898:335
Calonotus auratus.- Zerny, 1912:98.- Draudt, 1916:109 (as *C. aurata*) [invalid
    emendation]
Macrocneme thyridia.- Forbes (*nec* Hampson, 1898) 1939:130 [misidentification]

A widespread and variable species occurring from Mexico to Brazil. While often sympatric with other *Macrocneme*, it is most closely allied to *M. thyra*, with which it shares parallel geographic variation and a similar phenotype except for always being less white on the underside.

Since the type specimen of *lades* is lost, the following descriptions are based on a neotype male (Fig. 213) and associated female (Fig. 214) from Régina, French Guiana. The specimens were selected for their resemblance, especially of the female, to the original description and figure by Cramer. The type locality only approximates the original locality (Surinam) of *lades*, but the Guiana association is restrictive enough to assume that the same species is being treated.

MALE. *Head*: blackish brown, vertex and occiput with green metallic scales; labial palpi not reaching base of antennae. *Thorax*: blackish brown, iridescence bronzy green, metallic scales of disc not obscured by overlying hair-like scales; scattered metallic scales along hind margin of patagia; single, large metallic patch along anterior margin of tegulae; underside of tegulae white but scarcely visible from above along mesal margin; pectus hairy, blackish brown, with usual white markings including spot on propleuron; legs bronzy green with metalic scales in femoral grooves and mixed with some white on forecoxae; white spots on mid and hindcoxae and on tips of metatarsi. *Forewing*: brown, paler toward apex; bronze green spot at base; oblique black streak from basal angle to cell followed by medial, transverse green fascia with dark veins, distal margin of fascia not reaching end of cell; underside brown with basal third metallic green except where wings overlap; retinaculum white. *Hindwing*: brown, iridescent scaling absent; underside similarly brown, with basal half metallic green except in anal fold. *Abdomen*: iridescent green, mid-venter brown with mesal series of large white spots diminishing in size caudally. *Genitalia*: as in Figs. 47-53 (drawn from RED preps. 39217, USNM et al.; 16n); dorsal processes (claspers) of valvae symmetrical with inner surfaces at tips slightly concave; uncus rounded on dorsum, with lateral margins as short, subequal flanges; juxta extends beyond base of ventral processes, apex cut away at right and tip at left with row of short spines (cf. Figs. 49 and 52 for variation); aedeagus with two stout, unequal spines, the longer on the left, parallel to dorsum and tip curved outward, right spine short and raised at right angle to dorsum.

FEMALE. Almost identical to male, except iridescence of forewing restricted to single bronze green spot at base; black of basal angle diffuse and not reaching end of cell; white on abdominal venter restricted to mesal series of small, equal spots. *Genitalia*: as in Fig. 148-149 (drawn from RED prep. 255, UCB; 3n); sternite VII subquadrate, skewed slightly to left, with bottom right corner truncate, caudal margin entire, intersegmental membrane with thin, sclerotized band continuous at left with encircling pleurite and at right with sclerotized pocket consisting of two depressions, the more mesad deepest; sclerite of lamella postvaginalis oblong, margins eroded, lower half decurved at ostial opening; ductus bursae and stalk of accessory bursa with concentric plicae; medial protuberance of sternite VIII strongly notched mesad and continuous with right fold of underlying pair; signa of two opposing scallop-shaped patches with recumbent spines and 2-3 unattached spines nearby.

INDIVIDUAL VARIATION. Length of forewing: male, 14.5-17.0 mm; female, 15.7-17.8 mm. Wing ground color varies from blackish brown to brown, with tips occasionally non-pallescent; forewing iridescence varying from a single basal spot to a medial transverse fascia; hindwings normally devoid of iridescence on upperside but occasionally iridescent patch occurring below cell basad to Cu_{1+2}; oblique black streak from basal angle always present, but variable in width and length; wing iridescence either bronzy green, green, blue-green, or blue; color usually consistent within a population but occasionally mixed, especially blue with green; abdominal iridescence either bronzy green, green, or blue-green; color not necessarily consistent with that found on wings; thoracic pectus and abdominal venter whiter in males than in females; occasionally reduced white markings occur in males, especially on abdomen; length and shape of juxta varying slightly (cf. Figs. 49 and 52), but tip always extending above base of ventral processes; surface of left aedeagal spine either smooth or slightly spinose.

GEOGRAPHICAL VARIATION. Mexican and Central American specimens of *lades* have a blackish brown ground color on the wings and an iridescence that is either blue or sometimes green but never brassy. There is seldom any reduction in wing iridescence and the white markings, especially on the abdominal venter, are a similar series of spots.

The Colombian and western Venezuelan populations show only minor differences from their Mexican and Central American counterparts. The wing ground color is similarly blackish brown, but the iridescence may be either green or blue or a mixture of the two colors. The median transverse fascia of the forewing may be reduced considerably but is always present. The abdomen is usually green, but occasionally there is a mixture with blue. The amount of white on the abdominal venter varies among males, with some showing large spots basally and others having a series of small spots similar to the females.

In contrast, populations from eastern Venezuela and the Guianas have a phenotype that is distinctly brown. The apex of wings is slightly pallescent. Iridescence is usually bronze or sometimes green, but seldom blue. There is a large range in reduction of wing iridescence, with transverse median fascia sometimes completely absent. Abdominal markings usually separate males from females, with the base of the venter normally whiter in males. Walker's *Euchromia (Macrocneme) aurata* is a male of this brown phenotype. Its type locality is listed merely as 'Venezuela,' but this designation can be narrowed even further to 'lower region' of the Orinoco River, Venezuela, since this is a variant of *lades* that is easily separated geographically. Forbes (1939) obviously never saw the type specimen when he transferred *aurata* to *Macrocneme* from *Calonotos*. *Aurata* bears no resemblance to *semiviridis,* which he synonymized, or to *cabimensis,* which he mistakenly described and illustrated.

Although the distribution of *lades* is extensive and apparently continuous, its brown phenotype (=*aurata*) is distinguishable from other examples of *lades* by a geographic boundary that follows the outline for the Guiana Highlands, extending east of the Orinoco to French Guiana and south to the northeastern tributaries of the Amazon. The correlation between this browner phenotype and its distribution in the Guiana Highlands may be related, as Turner (1971) suggests, for the races of *Heliconius melpomene* and *erato,* to subspeciation occurring as a result of the isolation of populations during climatic cycles accompanying glacial periods. Presumably selection has acted similarly on the related species, *M. thyra,* because there are changes in appearance that clearly parallel those seen in *lades* when the two are sympatric in the Guianas. It appears that the two species are mimetically paired so closely that selective changes in one have been reflected in almost identical changes in the other. See *M. thyra* for further discussion of the similarities between these species.

TYPE DATA. *lades* Cramer: Surinam, type lost. Neotype male, **here designated,** French Guiana, Río Approuague at Régina, January 1-3, 1970, Pèré Y. Barbotin, USNM. Female, same data, USNM. ! *aurata* Walker: male holotype, Venezuela, BMNH.

The original description of *lades* might apply equally to occasional specimens of *thyridia* or *thyra* with the wing iridescence reduced to a single basal spot. All three species were described from Surinam without a specific locality. Although the male genitalia are distinctly different, each species has a variant phenotype which could be confused with the *lades* description. Under such conditions it seems appropriate in the interest of stability to have each species associated with a type specimen. Holotypes are extant for *thyridia* and *thyra* but for *lades* no type could be identified at the Rijksmuseum in Leiden where Cramer material would presumable be located. It is presumed lost and I have designated a neotype for *lades.* I suspect the original type was a female but have selected a male as neotype due to the greater certainty with which the male genitalia can be identified compared to the females. The different type locality is explained in the introductory remarks.

BIOLOGY. In Costa Rica, larvae of this species have been reared on *Mesechites trifida* [Apocynaceae]. The larvae eat mature leaves and are diurnal in activity. (Janzen, unpub.). See "Host Associations," p. 11 for further discussion.

ADULT HOST RECORDS. From Costa Rica, adults of *M. lades,* collected by P. Opler, visited inflorescences of the following plants:

Boraginaceae
    1. *Cordia curassavica* Roem. & Schult.
    2. *Cordia inermis* (Mill.) I.M. Johnston

Compositae
3. *Baltimora recta* L.
4. *Melanthera nivea* (L.) Small

GEOGRAPHICAL DISTRIBUTION. Very widespread. From Mexico through Central America to northern and central South America. Not known from Uruguay or Argentina.

FLIGHT PERIOD. Active throughout the year, since collection records are available for every month of the year at least in some part of its extensive range.

REMARKS. It appears from Cramer's description that *lades* was described from a female that had a simple bronze green spot at the base of the forewing. This wing pattern is not typical of the usual *lades* phenotype. In a series of *lades* specimens (n = 14) from Régina, French Guiana, one female (Fig. 214) is identical to Cramer's description but the others, both male and female, have a median fascia in addition to the basal spot. The iridescence in this band varies from moderate to light. A reduction in wing iridescence to a single spot is not unique to *lades*. Occasional specimens of *iole* Druce and *thyridia* Hampson show a similar reduction. A series of *iole* (n = 32) from San Vito, Costa Rica (UCB), has one female (Fig. 205) devoid of most iridescence, while the remaining specimens have the usual broad metallic band on the forewing. Similarly, one female of *thyridia* from Oyapok, French Guiana (RML), is identical to the Cramer illustration. The forewings are dark brown, and there is an absence of iridescence except for a basal spot. Other specimens of *thyridia* from eastern Venezuela (Río Mariusa, LACM) occasionally occur with the forewing iridescence reduced to a basal spot, but there is also a thin streak on the inner margin. The appearance is not identical to Cramer's description for *lades,* but similar enough to be confusing.

The most consistent character for separating *lades* from *iole* and *thyridia* is the structure of the male genitalia. *M. lades* has a scoop-like juxta that always extends beyond the base of the ventral processes of the valvae. The apex is rounded, spined on the left, and obliquely truncate on the right. In *thyra,* the juxta never reaches beyond the base of the ventral processes and the apex is straight. In *thyridia,* the juxta twists to the left, the apex is spined along its entire margin, and the right side is distinctly excavated. The three species show distinctive differences in the spines on the aedeagus (cf. Figs. 50, 57, and 70).

SPECIMENS EXAMINED (920): 526 males; 394 females. **MEXICO:** *Chiapas:* La Granja, AMNH; Huixtla, 20-25 mi N., CNC; Palenque, CNC; Portugal, 7 mi SE. Simojovel, UCB; Tapachula, Finca Violeta, SMM. *Guerrero:* Guerrero, BMNH. *Hidalgo:* Guerrero Mill, 9000 ft, BMNH. *Oaxaca:* Donaji, UCB; Palomares, MSU. *San Luis Potosí:* Tamazunchale, 2 & 25 mi N., 400', AMNH, UCB, USNM. *Tabasco:* Tabasco, BMNH; Teapa, BMNH. *Vera Cruz:* Acayucan, MSU; Acayucan, 30 mi S., UCD; Atoyac, BMNH, USNM; Coatepec, USNM; Córdova, BMNH, UCD, USNM; Cotaxtla Exp. Sta., Cotaxtla, UCB; Jalapa, AMNH, BMNH, USNM; Jicaltepec, CM; Lake Catemaco, CNC; Minatitlán, MSN; El Naranjo, UCB; Orizaba, AMNH, BMNH, PM; Paso San Juan, USNM; Ponpontler, AMNH; Vera Cruz, Presidio, AMNH; Salto Eyipantla [8 km S. San Andrés Tuxtla], UCB; San Andrés Tuxtla, MSU; Sta. Rosa, USNM; Sontecomapán, UCD; La Tinaje, 9 km S, UCB; Vera Cruz,

BMNH, PM. "Mexico" (Boucard, Genin, Ross, Salle, Salli), BMNH, PM, SMM, NRS. Isth. of Tehuantepec (Sumichrast), MCZ. **GUATEMALA**: *Alta Verapaz*: Panzós, Vera Paz, BMNH; Temahú, Vera Paz, BMNH. *Baja Verapaz*: Chejel (Schaus & Barnes), BMNH; Pancina [= Panimá], Vera Paz, BMNH; Purulhá [= Purulá], USNM. *El Progreso*: El Jícaro, Vera Paz, BMNH. *Escuintla*: Escuintla, BMNH; San José, CAS; Zapote [= El Zapote], BMNH. *Izabal*: Cayuga, BMNH, CM, USNM; Quiriguá, 240 ft, CM, CU, NRS, UCB. *Retalhuleu*: S. Sebastián, USNM. *Santa Rosa*: Guazacapán, BMNH. *Suchitepéquez*: Cuyotenango, USNM; San Isidro, 1600', BMNH. *Zacapa*: La Unión, LACM; Zacapa, USNM. "N. Guatemala" (Bedoc), PM; Guatemala, USNM; "Guatem. & C. Amer." (Hy. Edwards), AMNH. Not located: San Zornas, Atlantic coast, BMNH. **BELIZE**: *Belize*: Belize, (Rolfe, Johnson), BMNH, USNM; Manatee [Río], S. of Belize, BMNH. *Toledo*: Columbia [Río], BMNH; Punta Gorda, BMNH; Río Temas [= R. Temash], BMNH. "British Honduras" (White), BMNH. **HONDURAS**: *Atlántida*: La Ceiba, AMNH; Lancetilla, CU, MCZ; Tela, CU, CM. *Colón*: Puerto Castilla, MCZ. *Cortés*: La Cambre [= La Cumbre], 900-1000 m, BMNH; La Lima, USNM, USP; San Pedro Sula, AMNH, BMNH, NRS; Lago Yojoa, 8 km NE el Mochito, LACM. *Morazán*: [El] Zamorano, USNM. "Honduras", (Owen, Crowley, Schaus), AMNH, BMNH, CM, USNM; Siguatepeque, 6 km N, UCB; Taulabe, 6 km W, UCB. **NICARAGUA**: *Managua*: Los Nubes, 1300', AMNH. "Nicaragua" (DeLattre), BMNH. **COSTA RICA**: *Alajuela*: San Mateo, USNM; 2 km N. Tilarán, UCB. *Cartago*: Cache [= Cachí], BMNH; Juan Viñas, 2500-3500', BMNH, CM, USNM; Orosi, 1200 m, BMNH, PM; Sitio [de Avance], CM, USNM; Turrialba, CAS. *Guanacaste*: Bebedero, BMNH; Taboga, BMNH. *Limón*: Bataan [Batán], USNM; Colombiana, 300 ft, BMNH; Guápiles, CM; Hamburg Farm [bordering Reventazón River on Ramal Montecristo, elv. 10 m], MCZ. La Lola [Experiment Station, elv. 50 m], USNM; Puerto Limón, USNM. *Puntarenas*: 2 mi NE. Corredor, LACM; Golfo Dulce, [Pto.] Jiménez, VM; 1.8 mi W. Rincón, Osa Peninsla, LACM. *San José*: La Uruca, nr. San José, 1100 m, USNM; La Uruca, PM; San José, BMNH, PM, VM. "Costa Rica" (Boucard, Underwood, Serre, Biolley), BMNH, CM, PM. **PANAMA**: *Canal Zone*: Ancón, AMNH; Balboa, AMNH; Barro Colorado, AMNH, CAS, CU, MCZ; Empire, AMNH; Madden Dam, BMNH, MCZ; Canal Zone, BMNH. *Chiriquí*: Rovira, 1900', SE. Potrerillos Arriba, USNM; Volcan de Chiriquí, 2000-3000', BMNH; Chiriquí, AMNH, BMNH, CAS, VM. *Colón*: Colón, BMNH. *Panamá*: Las Cumbres, BMNH; Lino, 800 m, PM; Pacora, AMNH; Taboga Is., MCZ. *Veraguas*: "Veragua" [= Veraguas], BMNM. "Panama" (Ellison, Walker, Schaus & Barnes), AMNH, BMNH, USNM; Isthmus of Panamá, (Pemberton), BMNH. **COLOMBIA**: *Antioqia*: Puerto Berrio, BMNH. *Boyacá*: Muzo, R. Cantinero, BMNH. *Caldas*: La Dorada, AMNH, USNM. *César*: Lake Sapatoza Region, Chirigua[ná] Dist., BMNH. *Chocó*: Costa de Pacifico, Solano, SMM. *Cundinamarca*: Cananche, BMNH; Medina, 500 m, USNM; Monterredondo, 1310 m, 1420 m, SMM. *Magdalena*: Cacagualito, 1500', BMNH; Don Amo, 4000', BMNH; El Banco, Magdalena Valley, BMNH; Minca, 2000', BMNH; Onaca, Sta. Marta, 2000', BMNH; Río Frío, BMNH; Santa Marta, CAS, VM; Valparaiso, 4000', BMNH. *Meta*: Río Guayuriba, SMM; Río Negro, E. Colombia, 800m, BMNH; Villavicencio, E. Colombia, 400 m, PM, VM. *Santander del Sur*: Río Carare, 1000 m, SMM. *Tolima*: Honda, BMNH. *Valle*: Palmira, Valle de Cauca, USNM; Río Aguacatal, W.

Cordillera, 1600' & 2000 m, USNM, VM. "Colombia" (Marloff), CM. **VENEZUELA**: *Amazonas*: Carmen, USNM; Samariapo, USNM. *Aragua*: Cagua, 450 m, UCV; La Isleta, Choroní, 200 m, UCV; Maracay, SMM, UCV; Rancho Grande, 1100 m, SMM, USNM, UCV. *Barinas*: Campamento Cachicamos, Reserva Forestal Caparo, 100 m, UCV. *Bolívar*: Ciudad Bolívar, BMNH; El Barroso [Hato], Río Matú, [near Río Cuchivero], UCV; El Bochinche, Reserva Forestal Imataca, UCV; El Dorado-Sta. Elena, Km 107, 520 m, UCV; El Hormiguero, Meseta de Nuria, 500 m, UCV; Suapure, Caura Rive, CU. *Carabobo*: Las Quiguas nr. San Esteban, BMNH, CM, USNM; Valencia, BMNH; Río Borburata, E. Carabobo, 250 m, UCV; San Esteban nr. Puerto Cabello, BMNH. *Delta Amacuro*: La Horqueta 0-100 m, UCV. *Distrito Federal*: Caracas, BMNH. *Falcón*: Curimagua, UCV; Palma Sola, BMNH. *Lara*: Sanare, 1350 m, UCV. *Mérida*: Mérida, PM, USNM. *Miranda*: Guatopo, 400 m, UCV. *Monagas*: Caripe, 850 m, UCV; Jusepín, UCV; Río Morichal Largo [nr. bridge], UCV. *Tachira*: carretera La Fría-Coloncito, Km 2, orilla Río Oropé, 90 m, UCV. *Trujillo*: Cuicas, 1032 m, UCV; Trujillo, UCV; Valera, USNM. *Yaracuy*: Aroa, BMNH, UCV, USNM; Yumare, UCV. *Zúlia*: El Tucuco, 420 m, UCV; orillas Río Escalante á 30 km de Santa Cruz, UCV. "Venezuela," BMNH, USNM. **TRINIDAD**: *St. Andrew*: Manzanilla, OX. *St. George East*: Caparo, BMNH. *St. George West*: Ariapite, BMNH; Port of Spain, BMNH. *St. Patrick*: Cedros, USNM. "Trinidad," BMNH. **GUYANA**: *Mazaruni-Potaro*: Potaro, BMNH; Roraima, BMNH; Tumatumari, Rio Potaro, USNM; Plantain Island, Essequibo River, USNM. *West Demarara-Essequibo Coast*: Parika, BMNH. **SURINAM**: *Marowijne*: Aroewarwa Kreek, Maroewym Valley, BMNH. *Paramaribo*: Paramaribo, CU, BMNH. "Surinam," VM. **FRENCH GUIANA**: *Guyane*: Cayenne, BMNH, CM, PM, OX, USNM; Goebert, Maroni, BMNH; Nouveau Chantier, lower Maroni, USNM; Pied Saut, Oyapok R., CM; Rio Approuague at Régina, USNM; Roura, USNM; St. Jean du Maroni, USNM; St. Laurent, Maroni, USNM, VM. **ECUADOR**: *Bolívar*: Balzapamba, BMNH. *Esmeraldas*: San Mateo, SMM. *Los Ríos*: Pichilingue, 100 m, UCV. *Manabí*: Cojimies, CNC. *Napo*: Puerto Napo, 600 m, BMNH; Río Anzu [= Río Ansupí], IML; Sarzayacu [= Zatzayacu], 700 m, BMNH. *Pastaza*: Alpayacu, Río Pastaza, E. Ecuador, 3600', BMNH. *Tungurahua*: Ambato, BMNH; Río Topo, 1200 m, VM; La Merced, Río Pastaza below Baños, 4000', USNM. "Ecuador," BMNH. **PERU**: *Cajamarca*: Charape [= Charapi] River, Tabaconas, 4000', BMNH; E. slope Río Tabaconas, Charape, 4000', OX. *Cuzco*: Quincemil, 780 m, CNC. *Junín*: El Campamiento, Coloñia Perené, CU; Valle Chanchamayo, 800 m, IML. *Loreto*: Pucallpa, MSU. *San Martín*: Mayobamba, BMNH. Not located: Ygarapé, Upper Amazons, BMNH. **BOLIVIA**: *Cochabamba*: Chapare, 400 m, SMM; Chapare [upper region] Río Chipiriri, 400 m, SMM. **PARAGUAY**: "Paraguay," USNM; Río Paraguay, SMM. **BRAZIL**: *Amazonas*: Fonte Boa, BMNH; Hyutanahã [= Huitanaã], Rio Purús, CM; Nova Olinda, Rio Purús, CM; Rio Purús, NRS; São Paulo de Olivença, BMNH, USNM; Teffe [Tefé], CM, SW. *Mato Grosso*: Buriti, Chapada dos Guimarães, CNC; Salobra, USP. *Pará*: Altamira, Rio Xingú, USNM; Belém, MCZ, VM; Pará, BMNH; Ponte Nova, Rio Xingú, BMNH, USNM; Sesta, Lower Amazon, BMNH; Taperinha b. Santarém, VM, USNM. *Rio de Janeiro*: "Rio," BMNH, RML. *Rondônia*: Pôrto Velho, Rio Madeira, NRS. *São Paulo*: Ilha Seca, USP.

Macrocneme thyra Möschler

(Figs. 54-57, 150-151, 190, 198, 215; Map 15)

Macrocneme thyra Möschler, 1883:334, pl. 18, fig. 24
Macrocneme affinis Klages, 1906:539.- Hampson, 1914:206 [as synonym of *M. chrysitis*].- Draudt, 1917:204.- Zerny, 1931b:241 [NEW SYNONYMY]
Calonotos chlorota Dognin, 1914:Fasc. 7:8.- Hampson, 1914:215.- Draudt, 1917:205 [NEW SYNONYMY]
Macrocneme thyra boliviana Draudt, 1916:104 [NEW SYNONYMY]
Macrocneme thyra intacta Draudt, 1916:104.- Fleming, 1957:124.- Beebe & Kenedy, l957: 151 (behavior) [NEW SYNONYMY]
Macrocneme albiventer Dognin, 1923:Fasc. 23:2.- Forbes, 1939:129 [synonymy].- Fleming, 1957:124
Macrocneme chrysitis affinis.- Zerny, 1931a:13
Macrocneme thyra thyra.- Fleming, 1957:124

This is a widespread and extremely variable species, which accounts for much of its synonymy. By distribution and variation *M. thyra* most closely parallels *lades*, with which it is often sympatric. The two are easily confused, but *thyra* can always be distinguishd by the invariably whiter underside, the juxta truncated below the lower processes of the valvae, and a spiculate surface on the left spine of the aedeagus.

The following description is taken from a Guianan specimen (Wineperu, Guyana) which closely resembles the Möschler type from Surinam. Because the phenotype is variable, especially in color and pattern of the iridescence, not all individuals will follow in detail.

MALE. *Head*: blackish brown; white spots below antennae large and approximate on frons; white irrorations on vertex; labial palpi extending to base of antennae, front of II white with scattered white on III. *Thorax*: disc golden-green, margined along sutures with blackish brown hair-like scales; metallic green spots on patagia large and laterad; metallic green streak extending along fore and mesal margins of tegulae; underside of tegulae white, but visible from above only along mesal margins; pectus mostly white, with some blackish brown scales beneath wings and interspersed around coxae; legs blackish brown; coxae and femoral grooves with opposing margins on fore and mid tibiae white; inner mesal margins of fore and mid femora tan; hind tarsi white-tipped. *Forewing*: ground color blackish brown, basal half metallic green except along veins; small oblique black streak from basal angle; metallic scales not reaching end of cell, appearing more concentrated below cell and along inner margin; underside blackish brown with basal half metallic green; white streak along $A_1 + A_2$, tan where wing overlaps, retinaculum white. *Hindwing*: blackish brown with no iridescent scales; costal margin tan, with a diffuse white spot basad; underside mostly metallic green except at outer margin. *Abdomen*: metallic green, basal tergite black with metallic spot between subdorsal white spots; underside entirely white. *Genitalia*: as in Figs. 54-57 (drawn from RED prep. 39214, USNM, 18n); claspers of valvae slightly asymmetrical, with left arm somewhat longer and more curved than right arm, tip of left arm with dorsal edge more pronounced and emarginate than corresponding edge on right tip; dorsum of uncus slightly convex, with medial sulcus and lateral flanges; juxta truncate below ventral process of valvae, small spot of spines in membrane at left; dorsum of aedeagus with two strong spines of unequal length; left spine spiculate on ventral surface except at tip.

FEMALE. Essentially identical to the male except: *Head*: white absent from II and III of labial palpi. *Thorax*: white on underside restricted to small mesal dots, making underside appear darker than in males; pectus blackish brown, with white absent except as spot on pro and metapleuron; legs brownish black, with white spots only on forecoxae, trochanters, and hind tarsi; tan scales absent from mesal margin of fore and mid femora. *Wings*: ground color of wings slightly darker, approaching brownish black; iridescent scales sometimes sparse, blue-green instead of bright green; white streak absent along $A_1 + A_2$ on underside of forewing. *Abdomen*: white on abdomen restricted to medial row of small spots and to subventral pair on basal sternite. *Genitalia*: as in Figs. 150-151 (drawn from RED preps. 39213, USNM, and 39264, UCB; 13n); sternite VII shield-shaped, with posterior margin irregular, notched mesad, anterior margin with apex broad, skewed left, and two shallow, intersegmental depressions on right, one separate and the other continuous from margin of encircling pleurite; lamella antevagnalis with scattered longitudinal plicae; sclerite of postvaginalis roughly oval, decurved at middle where ostium originates; ductus bursae, corpus bursae, and stalk of accessory bursa with concentric plicae; stalk of accessory comparatively short, with bursa not extending beyond corpus bursae; signa of two opposing scallop-shaped patches with recumbent spines.

INDIVIDUAL VARIATION. Length of forewing: male, 15.8-18.2 mm; female, 15.8-18.0 mm. In all the descriptions of *thyra* or of one of its forms, the ground color of the wings and thorax is given as black. The color is actually brownish black, with certain populations, especially from the Guianas and eastern Venezuela, having a distinctly browner hue. This may in part be due to age, but fresh specimens as well as those in old collections can accurately be predicted as originating from these areas without reference to their collection labels. Individuals from any one locality normally have the same ground color, and the sexes are identical.

The only truly black scales on the wings form a streak at the base of the forewing. They are present either in a horizontal streak below the cell or as an oblique streak extending through the cell to the costa, or they are absent. Both *M. intacta* and *boliviana* were distinguished as subspecies partly on the basis of the forewings lacking these black scales, which made the iridescence appear entire. An examination of a series of specimens from localities as far apart as Trinidad and Bolivia shows that black scales are normally present and interrupt the iridescence except in a few specimens. The character is highly variable, and its absence in individuals is within the usual range of variation found in *Macrocneme*. Little taxonomic value should be attached to it.

The color of the wing iridescence is usually consistent within a population, but may vary between populations from golden-green, to green, to blue-green, to deep blue. Where metallic scales extend to the termen in the forewing (=*boliviana*), the color varies from dull blue to blue-green in the specimens from Bolivia, while specimens from Tabaconas (Perú) are bright green, and those from Guatopo

National Park (Venezuela) dark blue. Sometimes, within a single population, combinations of colors occur, as in three specimens with similar data from Simla, Arima Valley, Trinidad. In one both wings are green, in a second they are blue, and in the third the forewing is green and the hindwing blue.

GEOGRAPHICAL VARIATION. Although most species of *Macrocneme* show some form of geographical variation, in *thyra* the problem is complicated by a very wide distribution that has numerous geographic components, each appearing slightly different. The geographic limits for these are not always clear, since individuals even within a colony vary. Invariably the ranges overlap. Possible combinations in independently varying characters are so numerous that any attempt to describe forms is meaningless. Nevertheless some trends exist, and a few local populations are consistent enough to be recognized geographically. The typical Guianan form of *thyra*, as represented by the male holotype from Surinam, or *albiventer* Dognin from French Guiana, is wholly white on the pectus and venter, and the forewing underside is streaked with white on the inner margin. In other parts of the range, the venter is usually less white, especially toward the apex. Occasional specimens from widely disjunct populations (Pará or Mirapinima, Brazil; Quincemil, Perú; or Macas, Ecuador) may be as white as the Guiana form or even whiter (Perené, Perú; Port of Spain, Trinidad) if the inner margin on the forewing and the hindwing costa become extensively white basally. The usual pattern is for the three basal abdominal segments to be entirely white and the remaining segments to vary from mostly white to having a series of small mesal dots. Specimens that are profusely white on the underside (cf. Perené and Trinidad) tend also to be more extensively white on the legs and labial palpi.

The amount of white on the wing underside can vary considerably among individuals from one geographic area. In Trinidad a specimen from Port of Spain is extensively white at the inner margin of the forewing and at the hindwing base, while a specimen from Simla shows almost no white in these areas. Hampson (1898) considered the latter variant to be a subspecies of *thyra*, and Draudt (1916) named it *intacta*. I see no evidence of geographic isolation in this character. Similar variation is also found in mainland forms from the Guianas to Perú. The genitalia in all forms are essentially identical.

The most prevalent interpopulational variation in *thyra* occurs in color and pattern of iridescence. The usual pattern on the forewing is for the basal half to be covered with metallic green scales which do not extend beyond the discocellular, and a variable black streak extending distad from the basal angle. The iridescence may be uninterrupted, but generally the veins appear as thin dark lines. In some specimens the iridescence is reduced to a basal spot and a streak along the inner margin. In others the distal margin of the iridescene is lost in a general metallic suffusion extending beyond the cell towards the termen. The color is usually blue or blue-green. Hampson (1898) considered this variant distinct enough to be a subspecies, and Draudt (1916) subsequently named it *boliviana*. As the name suggests, the form commonly occurs in Bolivia (Coroico; Río Zongo), but it is not restricted to this area, occurring also in Perú (Tabaconas) and Venezuela (Guatopo National Park; Río Morichal Largo). A difference in the Venezuelan populations is that the suffusion is found only among females. This dimorphism is weak and possibly an artifact of the small sample size, since other females in the same population appear identical to the males.

Most examples of *thyra* have a patch of metallic scales in the discal area of the hindwings, but in specimens from the Guianas and eastern Venezuela this iridescence is absent, or occasionally present only as a few scattered scales. In one specimen from El Dorado (Venezuela), the hindwings are fully iridescent, while among specimens from Maracay (Venezuela) the metallic patch is either present or absent on the hindwing. Specimens from Panama (Chiriquí; Rovira) and Costa Rica show no iridescence in the hindwing, yet the populations in Mexico (Veracruz) clearly possess it.

None of the authors who named forms of *thyra* mention the iridescent striae on the abdomen. These are present in numerous specimens. The dorsum can be either iridescent green or have the metallic scales restricted to three longitudinal striae. There are two sublateral lines extending the length of the abdomen which are broader than the thin mid-dorsal line that may obsolesce beyond the middle. In the females the abdominal iridescence is often duller and darker than in males, making the striae appear faint or absent. These striae are particularly notable in the Trinidad populations (=*intacta*) and in individuals from the Guianas and Venezuela. The Andean, Central American, and Mexican populations are usually entirely iridescent with the striae not evident.

Besides *boliviana* and *intacta*, there is a form from Ecuador (Paramba; Atacumes) and Colombia (Río Dagua; Medellín) in which the distal margin of the forewing iridescence is sharply delineated and bright green. It closely resembles examples of *semiviridis* from the same localities and could be confused, except that its hind tarsi are white-tipped. Other specimens from Colombia, although showing the same sharp delineation in the forewing iridescence, may be blue-green (Santa Cecilia), or the male may be blue and the female green (Albán).

Where *thyra* occurs sympatrically with *lades,* the two species share similar patterns in variation. Both are browner in ground color in the Orinoco and Guiana drainages than in other parts of their range. In the Guianas and eastern Venezuela, where *thyra* is heavily white on the underside, *lades* has the white on the venter attenuating caudally. In western Venezuela, Costa Rica, and Mexico, where *thyra* is white at the base of the venter and has mesal spots caudally, *lades* is always less white and usually only has a mesal series of spots. Sometimes the venters appear identical in these species, but *thyra* males can always be distinguished by the white scales on the palpi and on the inner margin of the forewing underside.

TYPE DATA. ! *thyra* Möschler: male holotype, interior of Surinam, MNHU. ! *affinis* Klages: female holotype, Suapure, Venezuela, Dec. 30, 1899, USNM cat. no. 8414, USNM. ! *chlorota* Dognin: male holotype, Yuntas [=Juntas], near Cali, Colombia, (Fassl), USNM. ! *boliviana* Draudt: male holotype, Río Songo [=R. Zongo], Bolivia, BMNH. ! *intacta* Draudt: male lectotype, **here designated**, Trinidad, (Caracciolo), BMNH. ! *albiventer* Dognin: male holotype, Nouveau Chantier, French Guiana, USNM.

BIOLOGY. The host-plant association for *thyra* is unknown. I attempted to rear the species on an artificial diet modified from Shorey and Hale (1965). The rearings were initiated in Venezuela in May 1975 by collecting gravid females at mercury vapor lights in El Guapo and at El Lucero in Guatopo National Park. Seven individuals successfully oviposited in plastic bags.

Egg: Near white, semi-spherical, shiny, smooth, 0.8 mm wide; chorion transparent and possessing tiny hexagonal reticulations over entire surface above substrate; eggs deposited singly or in small clusters in polyethylene bags; eggs not in contact with each other; head capsule visible through chorion before emergence; eclosion occurred 4-8 days after oviposition at 24°C; larvae consumed chorion upon emergence.

Larva: **Instar 1**: length 6.5 mm, head capsule light brown, width 0.42-0.50 mm; body integument white, except abdominal segments A1, A3, A5, A7 appearing as lightly sclerotized bands; setae black, plumose, arising from tiny chalazae; secondary setae absent. **Instar 2**: head capsule black, width 0.54-0.69 mm; body integument light gray; dorsomeson, subdorsal, and lateral areas with white longitudinal lines; sclerotized bands on abdomen darker and more prominent than previous instar; caudal margin on metathorax (T3) yellow; paired, long white setae on T3 and A8. **Instar 3**: head capsule width 0.81-1.0 mm; color pattern similar to second instar except caudal margin of abdominal segments A2, A4, A6, A8 yellow and sclerotized areas of verrucae iridescent steel blue. **Instar 4**: essentially as described for full-grown larva except smaller and verricules not developed on A1 and A7; head capsule width 1.08-1.39 mm. **Instar 5**: (Fig. 198) full-grown larva (ex RED rearing lot 9E75, Guatopo National Park, Miranda Venezuela); length 23 mm; head capsule black, width 1.54-1.81 mm; body integument gray, surface spinulate especially around bases of verrucae; setal tufts black, plumose; cervical shield dark brown; subdorsal and lateral areas with thin white line; metathorax and abdominal segments A1-A8 with dorsocaudal margin narrowly yellow; yellow pigmentation variable but often more extensive on segments T3, A2, A4, A6, and A8; dorsomeson with prominent white stripe; bases of verrucae, especially D and SD (T2-A9), L (T1 and A9), L3 (A1-A8), and SV (T1-T3) strongly sclerotized and iridescent blue; L1 and L2 enlarged and modified as black-tufted verricules on A1, similarly L1 on A7; verrucae D+SD and L of T3 and SD of A8 with long white seta, dark-tipped; extra-long black tufts projecting cephalad from verruca D+SD of T2; short white setae interspersed among longer black setae on verrucae SD and L3 of abdomen.

Larval Development: Average time for the most rapidly developing larvae (n=9) to reach 5th instar was 22 days (range, 16-28 days). For those larvae moulting to a 6th instar (n=22) the average time for the most rapidly developing individuals (n=7) was 28 days (range, 22-31 days). In one lot (13E75), additional moults produced 7th (n=6), 8th (n=5), and 9th (n=2) instar larvae. All succumbed before pupating. One larva (5th instar) succeeded in pupating after 42 days, but its emergence as an adult was only partially completed.

    The wide variance in number of instars and in the time necessary to complete larval development is presumably related to nutritional deficiencies in the diet. Five instars are assumed to be the normal number, since the only two larvae that spun cocoons were 5th instar. The mortality rate was highest among 4th and 5th instar larvae. The response for those larvae not succumbing at these stages was to continue moulting up to four additional times. Pupation was not attempted, indicating that dietary deficiencies probably interfered with the cues necessary for its initiation.

Pupa: 16 mm long, yellow to yellow-brown; appendages on pupal shell outlined in black; abdominal segments banded black along margins; cocoon light tan with the black plumose setae of larvae interwoven over exterior; pupa suspended in middle and visible through setae.

GEOGRAPHICAL DISTRIBUTION. Widespread. Commonly encountered in Panamá, Venezuela, and Trinidad, in the Guianas, on Grenada, and on the tributaries of the Upper Amazon in Ecuador, Perú, and Bolivia; scattered records in Brazil include sites on the Amazon River and from the southeast; the single record from Mexico is incongruous and not explained.

FLIGHT PERIOD. Collecting records show *thyra* to be active in every month of the year.

REMARKS. To define *thyra*, I have overlooked the plasticity of the phenotype and emphasized the non-varying structure of the male genitalia. I have considered all specimens with a short juxta squared off below the point of origin of the lower process of the valve to be *thyra*. Also, the inner margin of the left clasper of the valve is distinctly emarginate before the tip. By contrast, in the phenotypically similar and sympatric *lades*, the juxta, although incised on the right, always extends beyond the base of the lower process of the valve, and the emargination on the clasper is absent, the edge smooth to the tip.

The sexes are dimorphic in the white markings and sometimes in the pattern of wing iridescence. These differences are given under the description for the female.

I have difficulty distinguishing between females of *thyra* and *lades*. The caudal margin of sternite VII (Fig. 150) is variable. The mesal emargination is sometimes absent. The internal genitalia are so similar that positive identification is usually conjectural unless the sexes are collected together and the males identified.

All the names synonymized under *thyra* represent geographic variants, except *albiventer* which was applied to a specimen identical to the Möschler holotype. The placement of *affinis* Klages in synonymy here is questionable since the holotype is a female which I am unable to associate with a known male. Its former synonymy with *chrysitis* Guérin is inaccurate since *chrysitis* is restricted to Mexico and Guatemala. Also the hind tarsi of *chrysitis* are black, whereas in *affinis* they are white-tipped. In the original description Klages allied *affinis* with *thyridia* Hampson. The *affinis* holotype is similar enough to *thyridia* to suggest that it is in fact merely the unassociated female. On the other hand the color and pattern of iridescence on the forewing and abdomen also closely resemble *thyra* females from Guatopo National Park, Venezuela, where I have been able to associate the sexes. Since I know of no other method for distinguishing the females of these two species, I am emphasizing the resemblance of the *affinis* holotype to *thyra* by the present synonymy. Future biological information may alter this placement.

The name *chlorota* has been overlooked in the literature in reference to *Macrocneme* due to its placement by Dognin in *Calonotos*. It is clearly a *Macrocneme* and of the form of *thyra* from Colombia where the abdominal underside is less white than the Guianan form.

The names *boliviana* and *intacta* refer to geographic forms of *thyra* from Bolivia and Trinidad respectively. Their synonymy is discussed under "Geographical Variation".

SPECIMENS EXAMINED (565): 349 males, 216 females. **MEXICO**: *Veracruz:Vera Cruz, USNM.* **PANAMA**: *Bocas del Toro*: Bocasdelton [=Bocas del Toro], USNM. *Canal Zone*: Barro Colorado, MCZ; Frijoles, MCZ; Madden Dam, BMNH, MCZ. *Chiriquí*: Bogavo [=Bugaba], BMNH; Chiriquí, CAS, CU, USNM; Rovira, SE of Potrerillos Arriba, 1900 ft, USNM. *Panama*: Las Cumbres, MCZ. *Veraguas*: Veragua (sic), BMNH. "Isthmus of Panama," BMNH; Panama, USNM. **COLOMBIA:** *Antioquia:* Medellín, USNM; Medellín, La Estrella, 1700 m, SMM. *Boyacá*: Muzo nr. Bogotá, USNM; Muzo, Río Cantinero, 400-800 m, BMNH. *Risaralda*: Sta. Cecilia, CM. *Chocó*: Andagoya, Río Condoto, BMNH. *Cundinamarca*: region of Bogotá, BMNH; Finca San Pablo, 3 km N. Albán, AMNH; Cananche, BMNH; La Vega, E. of Bogotá, dry, 1900 m, BMNH; Medina, 500 m, PM; Monterredondo, SMM; Pacho, 2200 m, BMNH, PM; Santa Fé de Bogotá, USNM; Susumico [=Quebrada Susumuco], USNM. *Magdalena*: Don Amo, 2000-4000 ft, BMNH; Minca, 2000 ft, BMNH; Onaca, Sta. Marta, wet season, 2000-2500 ft, BMNH; San Pedro (de la Sierra), 1400 m, USNM. *Meta*: Río Guayuriba, SMM; Río Negro, 800 m, BMNH; Río Negro nr. Villavicencio, USNM; Villavicencio, USNM. *Putamayo*: La Caravina, PM; Guineo, Río Putamayo, PM; Mocoa, 530 m, PM. *Santander/Sur*: Río Carare, 900 m, SMM. *Tolima*: Honda, BMNH. *Valle*: Buenaventua, CAS; Cali, USNM; Cali, Cauca Valley, 3260 ft, AMNH; Juntas [Río Dagua, 1000 ft]. ? *Huila*: 50 mi SW St. Augustine [=San Agustín], 7000 ft, CU. "Colombia," BMNH, USNM; Interior of Colombia, BMNH. **VENEZUELA**: *Aragua*: El Castaño (Maracay), 500 m, UCV; Güiripa near San Casimiro, 780 m, UCV; La Isleta, Choroní, 700 m, UCV; Maracay, SMM; Pozo Diablo (Maracay), 500 m, UCV; Rancho Grande, 1100 m, CU, UCV. *Bolívar*: El Dorado-Sta. Elena, km 107, 520 m, UCV; La Vuelta, Río Caura, BMNH. *Carabobo*: Las Quiguas nr. San Esteban, BMNH, CM, USNM; Puerto Cabello, BMNH. *Distrito Federal*: Caracas, BMNH, USNM; Caracas, Berg Avila, SMM; Todosana (via Caruao), 0-20 m, UCV. *Lara*: Sanare, 350 m, UCV. *Mérida*: Mérida, BMNH, USNM. *Miranda*: Parque Nacional Guatopo (Agua Blanca, 500 m; [El Lucero] 24 km N. Altagracia de Orituco, 640 m; La Macanilla, 500 m), UCV. *Monagas*: El Guacharo, 3 km N., UCV; Jusepín, UCV; Río Morichal Largo (puente), UCV. *Táchira*: La Fría-Coloncito, km 2, orilla Río Orope, 90 m, UCV; La Morita, 300 m, UCV; San José de Navay, 225 m, UCV. *Trujillo*: Valera, USNM. *Yaracuy*: Aroa, USNM. "Venezuela," BMNH, PM. **TRINIDAD & TOBAGO**: *St. Andrew*: Guaico, 150 ft, OX. *St. George*: Ariapite Valley, BMNH; Simla, Arima Valley, 800-1200 ft, AMNH; Upper Arima Valley, 1100-1800 ft, CAS; Belmont, VM; Caparo Valley (Port-of-Spain), BMNH; Carenage, Heights of Aripo, CM; Guanapo Road, AMNH; Hololo Mt. Road, 2000 ft, OX; Maraval, OX; Mont Tuchuche, 1000 m, PM; Port-of-Spain, BMNH; St. Anns, Port-of-Spain, AMNH; St. Anns Valley, BMNH; St. Augustine, BMNH; "Trinidad," BMNH, CM, OX; Tobago, BMNH. **GUYANA**: *Mazaruni-Potaro*: West bank, Essequibo River, BMNH; Picrewana Island, 6 mi S. Wineperu, USNM; Bartica, BMNH."British

Guiana," BMNH. **SURINAM**: Onoribo, BMNH. **FRENCH GUIANA**: *Guyane*: Pied Saut, Oyapock River, CM; Macouria, 100 m, PM; St. Georges, Oyapock, PM. **GRENADA**: St. George's, BMNH, MCZ. **ECUADOR**: *Esmeraldas*: Atacumes, CAS. *Imbabura*: Paramba, BMNH (*in copulo* pair); Paramba, 3500 ft, BMNH; Río Mira, Paramba, OX. *Santiago-Zamora*: Macas, USNM. Ecuador, BMNH. **PERU**: *Cajamarca*: Río Tabaconas, 6000 ft, BMNH. *Cuzco*: Quincemil, 780 m, CNC. *Huánuco*: Cushi, 1900 m, BMNH; Palcazú, 235 m, BMNH. *Junín*: Chanchamayo, 1000-1500 m, BMNH; Pueblo Pardo, Col[onia] de Perené, CU; Río Perené, CU. *Lima*: Lima, USNM. *Pasco*: Oxapampa, CU. *Puno*: La Oroya[=Oroya], Río Inambari, 3100 ft, BMNH; La Unión, Río Huacamayo, 2000 ft, BMNH; Sto. Domingo, Carabaya, 6000 ft, BMNH; Tinguri [Hacienda], Carabaya, 3400 ft, BMNH. *San Martín*: Tarapoto, CU. **BOLIVIA**: *Chocabamba*: Charaplaya [=Charapaya], 1300 m, BMNH. *La Paz*: Coroico, 1800 m, BMNH; La Paz, 1000 m, BMNH; Río Songo [=R. Zongo], 750-800 m, MCZ, PM, VM; Yungas de Coroico, 1800 m, BMNH. *Santa Cruz*: Buenavista, 750 m, BMNH. **BRAZIL**: *Alagoas*: Maceo [?=Maceió], BMNH. *Amazonas*: Rio Negro, Mirapinima, CNC; Pebas, BMNH. *Minas Gerais*: Bello Horizonte, 3000 ft, BMNH. *Pará*: Benevides, CM; Mosqueiro, Rio de Pará, BMNH; Pará, BMNH; *Paraná*: Castro, BMNH. *Rio de Janeiro*: Rio de Janeiro, PM.

Macrocneme thyridia Hampson

(Figs. 67-70, 154-155, 189, 216-217; Map: 16)

Macrocneme thyridia Hampson, 1898:321, pl. XI, fig. 9
Macrocneme guyanensis Dognin, 1911:6.- [synonymy] Forbes, 1939:130
Macrocneme euphrasia Schaus, 1924: 10. - [synonymy] Forbes, 1939: 130

This species may be easily confused with *M. lades* and *thyra*. They occur sympatrically and are similarly variable in color and pattern of the wing iridescence. Usually the forewing in *thyridia* has a concentration of green scales (sometimes blue) at the base of the wing accompanied by a similar concentration on the inner margin. The only dependable method to diagnose *thyridia* is to examine the genitalia. The three-spined aedeagus is unique among males of *Macrocneme*, and the sclerotized ductus bursae with an attendant blind pocket from the dorsal wall is unique among females.

MALE. *Head*: brownish black; white irrorations occasionally on clypeus or at vertex; metallic green scales between antennae on occiput; labial palpi upturned, not reaching vertex; scaling of II smooth, occasionally with white irrorations on outer surface. *Thorax*: brownish black, disc bright brassy green; patagia with metallic green scales along hind margin; tegulae with metallic green scales along fore margin and as mesal streak, underside white with tips visible from mesal fringe; pectus with white spots on propleuron and metepisternum; legs brownish black to black; hind tarsi white-tipped. *Forewing*: dark brown to almost black, appearing to fade toward apex; anterior white spot extending under tegula to costal edge and sometimes visible from below; iridescent scales either brassy green, blue, or blue-green; pattern varies, but characteristically an ill-defined, oblique band of black scales separates two narrow metallic green streaks from basal angle to midpoint below cell and along inner margin to before tornus; occasionally brassy scales at base of costa or scattered along Sc margin; median area often with an irregular fascia of metallic blue or brassy green scales; fascia sometimes indistinguishable when a general suffusion of metallic scales covers the wing to subterminal area;occasionally metallic scales absent from forewing except basal and inner margin streaks. Underside: metallic green to blue scales forming wedge-shaped streak covering Sc, discocellular, and cubital areas to beyond middle; retinaculum white. *Hindwing*: brownish black, costal margin tawny; white scales scattered basally along Sc + R$_1$ + Rs; metallic scales often absent, or when present in suffused patch in discal and terminal areas. Underside: iridescent green streak along costa and into cell; basal white spot on Sc + Rs. *Abdomen*: iridescent green or blue-green above and dull greenish black below; dorsal iridescence often interrupted by faint metallic stria mid-dorsally and two broader striae laterally; white spotting as described for genus. *Genitalia*: as in Figs. 67-70 (drawn from RED preps. 39199 and 39200; UCV, 5n); dorsal processes (claspers) of valvae asymmetrical, left valva slightly longer and less curved than on right; tips rounded; inner surface slightly convex and margin smooth; ventral processes weakly sclerotized; apex of uncus rounded, with short lateral flanges not extending beyond width of uncus; juxta deeply incised at right, tip with comb-like margin of thickened setae; aedeagus with three prominent spines, two subequal in length, the third elongate with a row of short, thickened spiculae at base; vesica of two interconnected bursae with small accessory bursa behind (not seen in illustration).

FEMALE. Essentially as described for male. Undersurface of tegulae white as in male, but less visible on mesal margin. *Genitalia*: as in Figs. 154-155 (drawn from RED prep. no. 39232; USNM, 3n); sternite VII with apical margin skewed to right, broad, caudal margin with shallow emargination at middle; shallow, intersegmental pocket between sternites VI and VII; antrum and walls of ductus bursae sclerotized; small evaginated pocket from dorsal wall similarly sclerotized; sternite VIII with caudal margin emarginate medially, anterior apophyses as short arms from pocket-like lateral margins.

VARIATION. Length of forewing: male, 14.5-16.0 mm; female, 14.5-16.8 mm. Amount and color of wing iridescence widely variable. Color usually bright, almost brassy green, or less commonly blue-green, contrasts with blue or green iridescence of median fascia.

TYPE DATA. ! *thyridia* Hampson: female holotype, Surinam, BMNH. Male, Surinam, MNHU. ! *euphrasia* Schaus: male holotype, Potaro River, British Guiana, USNM. ! *guyanensis* Dognin: male holotype, Saint Laurent du Maroni, French Guiana, USNM.

BIOLOGY. Among females collected at light in Guatopo National Park at El Lucero were six specimens so similar in appearance that I initially considered them identical examples of *thyra*. When confined in plastic bags they oviposited in a similar pattern. No differences in size and color of the eggs were noted. The emerging larvae fed on an artificial diet modified from Shorey and Hale (1965). After one moult the larvae of one lot (12E75) appeared to have a different color pattern. Upon re-examining the voucher female, I found her to be slightly smaller than the other *thyra* females. Also, the black streak from the basal angle of the forewing was broader and the iridescence of the abdominal dorsum was confined to mid-dorsal and lateral striae rather than suffusing over the entire surface. The variation was not unusual for a species of *Macrocneme*, but the larvae from this female developed a strikingly different pattern, especially the contrasting yellow and gray bands on the cuticle and the verrucae with iridescent blue bases. A dissection of the genitalia showed a sclerotized ductus bursae from which a small blind pocket evaginated on the dorsal wall. These characters identified the species as *thyridia*.

Egg: Not distinguished from that described for *thyra q.v.*

Larva: **Instar 1**: length 4.30 ± mm, head capsule light yellow, width 0.42-0.46 mm; plumose setae from chalazae with all dorsal and subdorsal setae black and lateral and ventral setae colorless; D1 + D2 + SD1 from single chalaza on T2, T3, and A9; L1 and L2 separate on abdominal A1-A8 but united on A9; L3 absent; SV bisetose on thorax and abdomen except SV2 absent from A7 and A8. **Instar 2**: length 5.62 ± mm, head capsule light brown, width 0.54-0.58 mm; readily distinguished from earlier instar by presence of secondary setae from verrucae; L3 occurs as verruca on abdominal A1-A9; segments T3, A2, A4, and A8 yellow, with bases of verrucae lightly sclerotized; remaining segments light gray, with verrucae heavily sclerotized. **Instar 3**: length 8.93 ± mm, head capsule light brown, width 0.73-0.81 mm; similar to instar 2 with the addition of a dorsomedial and a subdorsal white stripe on segments T2 to A7; verrucae prominently sclerotized except on yellow segments. **Instar 4**: length 10.29 ± mm, head capsule dark blackish brown, width 1.15-1.19 mm; yellow and gray bands more intensely colored than in previous instars; abdominal A6 with yellow laterally; verrucae of yellow segments (T3, A2, A4, A8) unsclerotized, those on gray segments heavily sclerotized and iridescent steel blue; long white seta from verrucae D + SD and L of T3 and SD of A8; long black setal tufts projecting cephalad from verruca D + SD of T2. **Instar 5**: length 12.32 ± mm, head capsule width 1.35-1.54 mm. Identical to description for instar 6 except smaller. **Instar 6**: (Fig. 189) full-grown larva (ex RED rearing lot 12E75, Guatopo National Park at El Lucero, Venezuela); length 21 ± mm; head capsule entirely yellow, width 2.19 mm; body integument spinulate; setal tufts black, plumose; cervical shield light yellow; dorsomeson and supraspiracular areas with white stripe; segments alternate yellow and gray with bases of verrucae iridescent blue on gray segments;segments T1, T3, A4, and A8 entirely yellow, with verrucae bases non-iridescent; abdominal A2 yellow with only verruca SD iridescent blue, remaining segments gray with verrucae heavily sclerotized and iridescent steel blue; black-tufted verricules from L1 and L2 on abdominal A1 and from L9 on abdominal A7; extra-long black tufts projecting cephalad from verruca D + SD of thorax T2; single white seta from verrucae D + SD and L of thorax T3, and from SD and L of abdominal A8; short white setae mixed with longer black setae on abdominal L3; all setae white and short from verrucae SV and V of abdomen.

Larval Development: The time necessary for larvae of *thyridia* to reach maturity is only approximated here, due to high mortality in the single rearing lot and to the failure of any larvae to pupate. The number of instars to maturity was assumed to be 5 since the most rapidly developing larvae (n=8) went through 4 moults in an average of 19.5 days (range, 18-21 days) before showing any sign of impaired development. Four larvae went through an additional moult (6th instar) in an average time of 27.3 days (range, 25-31 days) before succumbing, most likely due to nutritional deficiencies of the artificial diet. Pupation was never initiated.

Pupa: Unkown.

GEOGRAPHICAL DISTRIBUTION. Western Colombia, northern and southeastern Venezuela, Trinidad, the Guianas, northern Brazil, and upper Amazon of Perú and Bolivia.

FLIGHT PERIOD. Adults have been collected in every month of the year.

REMARKS. Of the three species (*M. thyridia, lades,* and *thyra*) described from "Surinam," it is conceivable that *thyridia* is conspecific with *lades*. Type specimens for *thyridia* and *thyra* are extant, but that for *lades* has been lost. For stability I have applied the names *thyridia* and *thyra* to species for which type specimens exist, and have referred *lades* to a third species which I can distinguish by genitalia and for which I have designated a neotype.

The original description and plate figure for *M. lades* could equally apply to the variety of *thyridia* in which the forewing iridescence is reduced to a small basal spot, as in the specimen from Oyapock (RML). A similar reduction in wing iridescence occurs occasionally in *thyra* and in the allopatric species such as *iole* and *cabimensis*. The identity of *thyridia* is further complicated by the variability in the color of the wing iridescence. Usually the metallic scales are brassy green throughout, but in some specimens the iridescent scales in the median area are blue. The name *euphrasia* was applied by Schaus to a specimen with brassy green scales, while *guyanensis* was applied by Dognin to a specimen with blue scales.

Hampson based his description of *thyridia* on a male, but listed a female as 'type' when the name was published. Whether the sexes were properly associated is uncertain, but I have examined both specimens and, lacking any evidence to the contrary, have assumed the association to be correct. The male has the unmistakable genitalic characters illustrated here.

SPECIMENS EXAMINED (117): 67 males, 50 females. **COLOMBIA**: *Risaralda*: Cristalina, 1100 ft, BMNH. **VENEZUELA**: *Aragua*: Rancho Grande, UCD. *Bolívar*: Ciudad Bolívar, BMNH, USNM; Río Caura, Campamento Cecilia Magdalena MCZ; Río Caura, Suapure, CU. *Guárico*: Río Guariquito, 100 m, UCV. *Miranda*: Parque Nacional Guatopo, [El Lucero], 24 km N. Altagracia de Orituco, 640 m, May (Dietz), USNM. *Monagas*: 42 km SE Maturín, LACM; 60 km SE Maturín, LACM; Río Morichal Largo [at bridge], UCV. *Delta Amacuro*: 140 km NE Barrancas, Caño Mariusa, Orinoco Delta, LACM. "Venezuela," (Grisol), PM. **TRINIDAD**: *St. George West*: Ariapite Valley, BMNH; Four Amandes [=Fondes Amandes, St. Ann's], BMNH; Port of Spain, Emperor's Vale, OX; St. Ann's Valley, 400-600 ft, BMNH, USNM. "Trinidad," (Kayne), BMNH. **GUYANA**: *East Berbice-Courantyne*: Rio Berbice, BMNH. *Mazaruni-Potaro*: Bartica, BMNH; Essequibo River, west bank, BMNH; Tumatumari, Rio Potaro, AMNH. "British Guiana," (Bartlett; Rodway; Whiteby), BMNH. **SURINAM**: *Paramaribo*: Paramaribo, (Ellacombe), BMNH.

"Surinam," (Möschler), PM. **FRENCH GUIANA:** *Guyane*: Cayenne, BMNH, CM, PM; mouth of Kourou River, BMNH; Oyapock, RML; St. Laurent du Maroni, BMNH; Roueou, PM. "French Guiana," (Bar), BMNH. **PERU:** *San Martín*: Tarapoto, BMNH. **BOLIVIA:** *Santa Cruz*: Chiquitos, San Julian, eastern Bolivia, 400 m, BMNH. **BRAZIL:** *Amazonas*: Canadian Father's Pool, 2 km N Itacoatiara-Manaus Hwy., 11 km W Itacoatiara, CNC; Lago Janauacá, Rio Solimões, 40 km W Manaus, CNC; Igarapé Prêto, upper Amazon, BMNH; Manicoré, USNM; Rio Negro, Iguapo (permanently flooded forest) 5 km S. Rio Araras, CNC. *Mato Grosso*: Chapada near Cuyabá [=Cuiabá], CM. *Pará*: Pará [=Belém], BMNH; Villa Franca [=Vila Franca], BMNH.

Macrocneme coerulescens Dognin

(Figs. 62-66, 162-163, 191-194, 218-219; Map 17)

Macrocneme coerulescens Dognin, 1906: Fascicle 50:180
Macrocneme yepezi Förster, 1950:60 [NEW SYNONYMY]

This species occurs in three geographic forms that are similar enough in appearance to sympatric *M. thyra, lades,* and *durcata* that only an examination of the genitalia will assure a reliable identification. Unusual for species in the genus is the variation in hind tarsal scaling that creates populations with both black- and white-tipped individuals. Similarly, occasional variation in the right margin of the juxta is unknown in other *Macrocneme* except *semiviridis*.

In northern Venezuela, *M. coerulescens* occurs as a distinctive green form (=*yepezi* Förster; Fig. 219), and from the Andes of Venezuela and Colombia it resembles *lades* Cramer, except that the juxta is larger and more spinose at the apex. Dognin described this Andean form as *coerulescens*. Finally, from northern Perú *coerulescens* occurs as a large blue form (unnamed) with a disjunct range and a resemblance to *durcata*, except that the genitalia are indistinguishable from the Venezuelan *coerulescens*.

The following description is based on Dognin's concept of *coerulescens*. His holotype female and a plesiotype male from the same locality (Mérida, Venezuela) were the voucher specimens. The green and blue forms mentioned above are discussed under "Geographical Variation".

MALE. *Head*: brownish black, occiput with metallic blue spots; labial palpi not reaching base of antenna. *Thorax* (including legs): brownish black, metallic scales blue; disc iridescence clearly visible; metallic spots of patagia large, adjoining lateral white spots; tegulae with metallic scales along foremargin, white scales present on underside but not visible along mesal fringe; pectus brownish black, hirsute, single white spot on pleural segments; legs with metallic scales in coxal grooves, inner surface of coxae white, hind femora with thin white streak, hind tarsi white-tipped. *Forewing*: brownish black with basal half metallic green to blue-green, veins dark, iridescence not reaching end of cell or to tornus, short black streak below A_2.

Underside similar, except inner margin and retinaculum grayish brown. *Hindwing*: brownish black with discal area metallic blue, veins dark. Underside similar, but iridescence extends to inner margin, costa prominently metallic. *Abdomen*: dorsum iridescent, shiny green except basally, where blue scales of thoracic disc form small mesal spot; pleura with two white spots basally; venter with lateral margins metallic green, mesal series of white spots to sternite V plus additional pair on sternite I. *Genitalia*: as in Figs. 62-66 (drawn from RED prep. 39202 and 39226, USNM, 7n); dorsal processes (claspers) of valvae symmetrical, ventral processes subequal in length to claspers, slightly sclerotized on upper surface; dorsum of uncus round, smooth, lateral margins as vertical flanges; juxta 3.5-4.2 mm long, extending well above base of ventral process, apex broadly incised at right with mesal margin membranous, tips margined with 2-3 rows of uneven spines, diaphragma sclerotized at attachment of juxta; aedeagus with two spines, outer surface of left spine lightly spinose.

FEMALE. Essentially like male, except: Wing iridescence variable, green to blue-green proximally, bright blue distally on forewing; green scales appear concentrated in short streak below cell; black scales restricted to thin streak between A_1 and A_2. *Genitalia*: as in Figs. 162-163 (drawn from RED prep. 259, USNM, 8n); sternite VII triangulate with broad, truncated apex, caudal margin without mesal notch; intersegmental membrane VI-VII with broad sclerotized depressions from margins of encircling pleurites; small unattached sclerite mesad; lamella antevaginalis as narrow fold less than half width of outer fold of sternite VII; lamella postvaginalis roughly ovate, lying perpendicular to VIII, surface undulate, only slightly decurved at ostial opening; ductus bursae with sclerotized fold in wall at left; signum as for genus.

INDIVIDUAL VARIATION. Length of forewing: 16.8-18.6 mm. *Wings*: iridescence on upper surface of wings variable in color and pattern; metallic scales either blue, green, or a mixture of blue and green with one color dominating; iridescence of forewing typically occupying basal half of wing and extending only to end of cell; common variant has iridescence more diffuse and extending outward to tornus and occasionally to outer margin and apex; hindwing iridescence may extend to the outer margin, or be entirely absent; basal black streak of forewing variable, often thin and usually not touching cell, sometimes broad and reaching bottom of cell or extending into it. *Legs*: white of hind tarsi sometimes much reduced or absent. *Abdomen*: dorsum either entirely iridescent green or dark green with shiny iridescent scales restricted to lateral margins and to a thin mid-dorsal stripe; white spots of basal tergite and mesal series on venter variable in size from few scales to prominent tufts. *Genitalia*: incised right margin of juxta variable, occasional unique margin (Fig. 66) in otherwise normal series (ex Km 37, Barinitas); spines on aedeagus slightly variable in size, bases either parallel or right higher than left.

GEOGRAPHICAL VARIATION. The northern Venezuelan populations of *M. coerulescens*, described by Förster (1950) under the name *yepezi*, often have a wing iridescence that is distinctly green and diffuse to the tornus. The abdomens are dark green with thin iridescent stripes on the lateral margins and one mid-dorsally. The white spots on the basal tergite are reduced and the hind tarsi may have the usual white absent from the tips.

In the Andes of Venezuela and Colombia the green *yepezi* form (Fig. 219) is replaced by the typical *coerulescens* with the iridescence of the forewing restricted to the base and having prominent white spots on the basal tergite and hind tarsi. The iridescence may difuse to the tornus in some specimens, but the color is normally blue (Mérida) or blue-green (San Isidro). Occasional examples (Monterredondo, Curimagua, La Soledad) may be green like the *yepezi* form, but they are uncommon. The hindwings in the *yepezi* form always have metallic scales, but among Andean *coerulescens* these scales may be absent from some of the individuals in a population (Pregonero, Las Lajas, Monterredondo). The dark green abdomen with thin metallic stripes of the *yepezi* form is often seen in Andean *coerulescens,* but mixed in the same populations (Monterredondo) are individuals with abdomens so iridescent that the metallic stripes are obscure or not evident.

At the southern extremity of the *coerulescens* range in Perú the specimens from Perené, Pomacocha, and Tabaconas (n = 18) represent an anomalous form indistinguishable on the basis of the male genitalia from the northern *coerulescens,* but larger, more extensively iridescent blue on the wings, and disjunct in its distribution. The forewing measures 20.5-21.5 mm versus 16.8-18.6 mm for the northern populations of *coerulescens.* The iridescence reflects a deeper blue than in Andean specimens, and diffuses over more of the wings. The juxta is longer (4.2 mm versus 3.5 mm), which may only be due to the larger size of the specimens. The hind tarsi are prominently hirsute and only slightly white-tipped, and the spots of the basal tergite are small, like those seen in the *yepezi* form.

TYPE DATA. ! *coerulescens* Dognin: female holotype, Mérida, Venezuela, USNM Type# 30716, USNM. [Abdomen missing] ! *yepezi* Förster: male holotype, Caracas (Los Venados), Venezuela, June 1937, SMM.

BIOLOGY. Limited success at rearing the larvae of this species on an artificial diet confirmed my suspicion that the green form described as *yepezi* is probably a geographic variant of *coerulescens.* Four rearing lots were initiated from females collected at mercury vapor lights. Two *yepezi* females (RED Lots 8K74 and 15E75) were collected in the northern coastal range at Rancho Grande and at El Lucero in Guatopo National Park, respectively. One typical *coerulescens* female (Lot 2F75) was from Rancho Grande, while another (1F75) came from the Andes near the aqueduct tunnel of San Isidro, southwest of Barinitas.

The larvae from lots 15E75 and 1F75 are illustrated in Figs. 191 anr 192. While the ground color for the *yepezi* form (Fig. 191) is a darker gray than the typical *coerulescens* ((Fig. 192), other markings, including the band on the head capsule, the pattern of yellow spots, and the longitudinal lines on the body appear identical.

Egg: Near white, semi-spherical, shiny, smooth, 0.8-0.9 mm wide; chorion transparent and possessing tiny hexagonal reticulations over surface; eggs deposited in small clusters in polyethylene bags; eggs not touching each other, but arranged in parallel rows of 2-3 eggs per row; mandibles visible through chorion before emergence.

Larva: **Instar 1**: head capsule light brown, width 0.39-0.40 mm; setae plumose on small chalazae with D and SD black and L, SV, and V without pigment; body integument off-white. **Instar 2**: head capsule black, width 0.54-0.62 mm; numerous black secondary setae from small verrucae; body integument light gray; dorsum between T3 and A7 banded by yellow spots between verrucae D1 and D2 and D2 and SD. **Instar 3**: head capsule black, width 0.92-1.0 mm; similar to instar 2, except body integument darker gray and the transverse bands on dorsum T3-A7 more prominent; verrucae D and SD of segments A8-A10 iridescent blue; long, black setal tufts projecting cephalad from verrucae D+SD of T2; single white setae from T3 and A8. **Instar 4**: head capsule black, width 1.15-1.23 mm; similar to instar 3 but larger. **Instar 5**: as described for instar 6, except head capsule width 1.54-1.62 mm. This may be the final instar for larvae under natural conditions. Although additional moults occurred, the number and duration of the instars were too variable to determine whether they were normal in larval development or were the result of nutritional deficiencies, causing a delay in the onset of pupation. **Instar 6** (Fig. 191): full-grown larva (ex RED rearing lot 15E75, Guatopo National Park, Miranda, Venezuela; length 26 mm; head capsule with epicranium brown, black band across front, width 2.04 mm; body integument gray, surface finely spinulate; setal tufts black, plumose; cervical shield brown with mesal margins black; dorsal line light gray; three longitudinal dark gray lines in dorsal, subdorsal, and subspiracular areas from segment T2 to A9; dorsum from T3 to A7 appearing banded from verrucae D1, D2, and SD prominently sclerotized and spotted with yellow between D1 and D2 and between D2 and SD; single yellow spots caudad of verruca D+SD of T2 and ventrocaudad of SD on abdominal A8; verrucae D1 and D2 of A8, D+SD of T2 and A9, and dorsum of A10 irridescent steel blue; L1 and L2 of abdominal A1 enlarged and modified as black-tufted verricules, L1 of abdominal A7 similarly tufted; long white seta, dark-tipped, from verrucae D+SD and L of T3 and from D2 and SD of A8; extra-long black tufts projecting cephalad from verruca D+SD of T2; verruca L3 on abdom. A1-A8 with white setae subequal to and interspersed among black setae.

Larval Development: Eclosion occurred in 7-8 days at 24°C. Chorion was consumed immediately upon emergence. A few larvae always developed more rapidly than others in the same lot. The most vigorous (n=4) reached 5th instar between 17 and 40 days. For those larvae that moulted to a 6th instar (n=3), the development time varied between 26 and 49 days. Only two larvae pupated, one from a 6th instar (RED lot 8K74) and another from a 7th instar (RED lot 1F75). Total time from eclosion to completion of pupation for these larvae was 60 and 63 days, respectively. Only the pupa of lot 8K74 (Figs. 193-194) produced an adult female, 76 days after eclosion. The emerging female was smaller than normal and its wings were deformed.

The times for the development of *coerulescens* are only approximations of what might happen under natural conditions. None of the rearing lots produced a series of larvae from which a consistent pattern of development could be discerned. Deficiencies in the diet became particularly evident after the 4th moult. The larvae of one lot (2F75) stopped development and succumbed within a week. In three lots (8K74, 15E75, 1F75), the larvae moulted to a 6th instar and one larva pupated. The adult emerged deformed. In lot 1F75, five larvae moulted to a 7th instar, but the head capsule remained the same size as for the previous instar. Of these larvae, one was preserved, two died, one pupated (deformed), and one moulted a 7th time before eventually succumbing.

The final instar in many Lepidoptera is often signaled by striking changes in appearance following the final moult. In *coerulescens*, black-tufted verricules on segments A1 and A7 and a change in head-capsule color first occur after the 4th moult. As many as 3 additional moults occurred, but the appearance of these later instars remained identical to the larvae emerging from the 4th moult.

Pupa (Figs. 193-194): 13 mm long, dark brown; appearance very similar to that of *thyra*; appendages on pupal shell outlined in black; abdominal segments banded black along margins; cocoon white, oval, with black plumose setae interwoven over exterior; pupa visible through silken network of cocoon.

GEOGRAPHICAL DISTRIBUTION. Andean cordilleras of western and northern coastal Venezuela, central and eastern Colombia, and northern Perú. Elevations vary from sea level to 2000 m.

FLIGHT PERIOD. Active throughout the year. There are collection records for every month, with males far outnumbering females in their attraction to light.

REMARKS. A major difficulty in defining this species, aside from the geographic variation, was that the holotype was a unique female, and an associated male was not readily apparent. The genitalic preparation of the holotype abdomen was unfortunately lost by theft, leaving me only with a photograph of abdominal sternite VII. I originally considered *M. coerulescens* to be a synonym of *lades*, but a comparison of the 7th sternite photo of *coerulescens* with examples of *lades* indicated that two species were involved. Among a series from Monterredondo, one specimen was identical to the holotype from Mérida. It could be associated with males from the same series, and its 7th sternite appeared identical to that in the *coerulescens* photo. I have based the identity of the species on this comparison.

I have placed *yepezi* in synonymy because the populations to which the name refers appear to be merely geographic variants of *coerulescens*. The distribution of *yepezi* is somewhat restricted, but it overlaps with the Andean populations of *coerulescens*, and the genitalia appear identical. I have seen *yepezi* most often from the coastal ranges of northern Venezuela, particularly in the states of Aragua and Miranda. Occasionally the green form occurs further west in Carabobo and Lara, and rarely in the Andes, e.g., Barinas. Where the green *yepezi* form is sympatric with *durcata*, its identification is simplified by *durcata* being larger and having a bright blue wing iridescence. Förster compared *yepezi* to *cyanea* Schaus in his original description, but probably he referred to populations now assigned to *durcata*. *M. cyanea* is known only from southern Brazil, and *durcata* was undescribed at the time.

I have treated the blue Peruvian populations as the southernmost representatives of *coerulescens*. In fact, they may be a valid species. The size, color, and distribution are readily recognizable, but I cannot distinguish the genitalia. Since geographic variants are known for other species of *Macrocneme* (see *thyra, durcata, cupreipennis*), I feel the present treatment of the blue form as an unnamed race adequately recognizes the existence of the populations and avoids possible nomenclatorial problems until more material is available.

SPECIMENS EXAMINED (248): 206 males; 42 females. **COLOMBIA**: *Boyacá*: Muzo, 400-800 m, PM. *Cundinamarca*: Distrito Especial, Chocó, USNM; Bogotá,

BMNH; Monterredondo, 1420 m, SMM; Medina, E. Colombia, BMNH. *Meta*: Río Negro, E. Colombia, 800 m, BMNH; Villavicencio, USNM. *Santander/Sur*: Cúcuta, Venez. [sic], BMNH. *Valle*: Río Aguatal [=Río Aguacatal], USNM; Tocotá (Fassl), USNM; Pichindé, USNM. *Tolima*: Cañon del Monte Tolima, 1700 m, BMNH, PM; San Antonio, USNM. "Colombia," USNM; "N. Grenada," BMNH. Not located: Villa Elvira, 1600 m, USNM. **VENEZUELA**: *Aragua*: Choroní, 200 m, UCV; Güiripa nr. San Casimiro, 780 m, UCV; El Limon (Maracay), 450 m, UCV; Maracay, SMM; Rancho Grande, 1100 m, CAS, UCD, UCV, USNM, RML, SMM; La Victoria, 1700 m, SMM. *Barinas*: Barinitas (at Km 37), UCV, USNM; La Chimenea, 5 km S. La Soledad, 1500 m, UCV; La Soledad, 950 m, UCV; San Isidro, 14 km S. La Soledad, 1500 m, UCV. *Carabobo*: Las Quiguas, CM; Valencia, 1500-3000 ft, USNM. *Distrito Federal*: Caracas, BMNH; Berg Avila, SMM; Caracas, Los Venados, SMM; Río Catuche, PM; Macizo [=massif] Naiguatá, UCV. *Falcón*: Curimagua, 1120 m, UCV, USNM. *Guárico*: Río Guariquito [nr. Orinoco], 100 m, UCV. *Lara*: Terepaima, UCV. *Mérida*: Mérida, 1600 m, BMNH, NMG, UCV, USNM. *Miranda*: Parque Nacional Guatopo (Agua Blanca, 500 m; [El Lucero], 640 m; La Macanilla, 500 m), UCV; San Antonio, UCV; San Diego de los Altos, UCV. *Monagas*: 42 km SE Maturín, LACM. *Táchira*: La Grita, UCV; Las Lajas [via Betañia], 1700 m, UCV; Pregonero, UCV; Queniquea, 1600 m, UCV. "Venezuela," BMNH, CM. **PERU**: *Amazonas*: 5 km N. Pomacocha on road to Rioja, 2000 m, CAS. *Cajamarca*: Charape [=Charapi] & Tabaconas Rivers, 4000-6000 ft, BMNH, OX, PM, USNM. *Junín*: Pichis & Perené Vs. [Valleys ?], 2000-3000 ft, USNM.

Macrocneme ancaverdia Dietz, new species

(Figs. 75-78, 152-153, 221; Map 18)

This species is allied to *Macrocneme aurifera* Hampson and largely sympatric with it except in northern Venezuela and Trinidad. The male valvae are similarly acuminate at the tips but lack the inner thorn characteristic of *aurifera*. The juxta is large and spatulate, while in *aurifera* it is tiny and rod-like. The female sterigma lacks the paired intersegmental pockets found in *aurifera*, and the longitudinal fold of the lamella postvaginalis is absent. The hind tarsi are unusual among *Macrocneme* in that they can be either white or dark-tipped. Usually they are white, but in some individuals the white scaling is absent, or so reduced as to appear absent.

MALE. Length of forewing: 16.0 mm. *Head*: brownish black, two small metallic spots on occiput; labial palpi not reaching base of antennae. *Thorax* (including legs): brownish black, metallic markings blue; iridescence of disc somewhat obscured by overlying hair-like scales; patagia with comparatively large metallic spots adjoining lateral white spots; tegulae with metallic streak across anterior margin and partially mesad; underside with few white scales, not visible on mesal fringe; pectus without metallic scales, white spots of pro- and mesopleura small; metallic scales in coxal grooves; all tarsi metallic streaked on outer surface; hind tarsi white-tipped. *Forewing*: brownish black, basal half metallic blue, veins dark, iridescence not reaching end of cell, oblique black streak from basal angle to bottom of cell; underside similarly blue basad, inner margin light brownish black, not metallic, A_2 streaked white, retinaculum white. *Hindwing*: brownish black with small scattering of metallic blue in discal area; underside metallic only along costa and in cell. *Abdomen*: dorsum brownish black, lateral margins and thin mid-dorsal stria iridescent green, terminal segment blue-green; venter with lateral margins metallic, mesal series of small white dots diminishing in size distally; pleura with two unequal white spots at base, the larger on segment I. *Genitalia*: as in Figs. 75-78 (drawn from RED prep. 39206, UCV, 4n); dorsal process (claspers) of valvae slightly asymmetrical, left arm slightly longer and straighter than curved right arm, tips distinctly acuminate; uncus skews left when viewed dorsally, dorsum narrow with lateral margins as thin subhorizontal flanges; juxta large, extending well above base of ventral processes, apical margin entirely spined, incised at right, with lower end of incision terminating in sharp point; dorsum of aedeagus with two small spines pointing right, the left slightly stronger and sharper than right.

FEMALE. Similar to male except: Metallic pattern of wings occasionally differing; allotype female reflects common variant, with forewing iridescence suffusing beyond cell to tornus, not reaching outer margin; hindwing more heavily metallic in discal area and iridescence of underside extending below cell; terminal segment (VII) of pleura prominently white-spotted. *Genitalia*: as in Figs. 152-153 (drawn from allotype, RED prep. 277, BMNH, 5n); sterigma with sternite VII broadly U-shaped, apex cephalad and skewing left, lateral margins as asymmetrical flaps, caudal margin uneven but not emarginate medially; two sclerotized intersegmental depressions at right, one independent, the other continuous with margin of encircling pleurite; fold of lamella antevaginalis concave at middle; oval sclerite of lamella postvaginalis bent and narrowly inserted at ostium; sternite VIII with single thick plica to left of medial protuberance; ductus bursae with single thickened fold from dorsal wall (behind sclerite of lamella postvaginalis); signa comparatively large, scallop-shaped patches of recumbent spines.

VARIATION. Forewing length: males, 16.0-18.2 mm; females, 17.2-19.0 mm. Wing iridescence blue, green, or a mixture of blue and green; metallic green often concentrated as basal spot below cell when iridescence is mixed (cf. Río Negro, La Merced, or La Unión); iridescence of hindwing either sparse (cf. holotype) or as a prominent patch (cf. allotype); forewing black streak variable, either relatively broad or reduced to thin streak; hind tarsi usually white-tipped, but in some populations (Coroico, R. Zongo, Suapi, La Unión) white is greatly reduced or absent.

TYPE DATA. Holotype male: Lima, Perú (B.P. Clark, donor), USNM Type 73268, USNM. Allotype: La Merced, Perú, 2500', May/June '03 (Watkins & Tomlinson), BMNH. Paratypes (29): 16 males, 13 females. **VENEZUELA**: *Amazonas*: Yavita, February (Lichy), USNM. *Bolívar*: El Dorado-Sta. Elena, Km 125, 100 m, September (Rosales, Rodriguez, Gelbez), UCV; Río Matú [Hato] El Barroso [nr. Cuchivero], January (Salcedo), UCV. **COLOMBIA**: *Meta*: Río Negro, E. Colombia, 800 m (Fassl), BMNH; Villavicencio, 400 m (Fassl), PM. **ECUADOR**: *Bolívar*: Balzapamba (Haensch), BMNH. *Chimborazo*: Riobamba, 2798 m, BMNH. *Zamora-Chinchipe*: Zamora, 3-4000 ft (Baron), BMNH. **PERU**: *Huánuco*: Pozuzo, 800-1000 m (Hoffmann), BMNH. *Junín*: La Merced, 2000-3000 ft, May-August,

(Watkins; Watkins & Tomlinson), BMNH; Río Colorado, 2500 ft, July-August (Watkins & Tomlinson), BMNH. *Puno*: La Unión, Río Huacamayo, Carabaya, wet season, November, December, 2000 ft (Ockenden), BMNH; San Gabán [=Lanlacuni Bajo on Río Sangaban], 2500 ft, March-April, BMNH. *San Martín*: Tarapoto (de Mathan), BMNH. **BOLIVIA**: *La Paz*: Coroico, 1400 m, 1800 m (Fassl), VM, USNM; Río Songo [=R. Zongo], 750-800 m (Garlepp; Fassl), PM, USNM, VM; Suapi (Garlepp), USNM.

BIOLOGY. No information.

GEOGRAPHICAL DISTRIBUTION. Eastern regions of Venezuela, Colombia, Ecuador, Perú, and Bolivia.

FLIGHT PERIOD. Adult collection records are available for all months except May, June, and October.

ETYMOLOGY. The specific epithet, derived from the Spanish *anca* = ankle and *verde* = green, describes the iridescence sometimes occurring on the legs.

REMARKS. Apparently *M. ancaverdia* and *aurifera* occur in different frequencies in Perú. A series which appears homogeneous may contain both species, but the sexes may not be evenly distributed. Although the female type of *aurifera* was described from La Merced, Perú, males of the species have yet to be seen from Perú. On the other hand, both sexes of *M. ancaverdia* have been taken at La Merced.

Macrocneme aurifera Hampson

(Figs. 71-74, 158-159, 220; Map 19)

Macrocneme aurifera Hampson, 1914: 204, pl. XI, fig. 2; Draudt, 1916: 204
Macrocneme spinivalva Fleming, 1957: 125-126, pl. I, fig. 4; pl. II, fig. 4 [NEW SYNONYMY]

This species is only sporadically collected, usually in the company of other *Macrocneme*. The adults resemble *M. thyridia, thyra,* and *lades* in usually having a concentration of metallic green scales at the base of the forewing and in their similar distributions. Both sexes have white scales along the anterior margin of the patagia connecting the dorsal and lateral spots. The males are unique for the tiny, rod-like juxta and the thorn-like projection from the inner margin of the claspers. The females have a diagnostic longitudinal fold in the sclerite of the lamella postvaginalis, and the intersegmental membrane between sternites VI and VII has two approximate pockets with a tiny cuneiform sclerite intervening. Females of *durcata* have similar pockets, but farther apart and with a correspondingly larger sclerite medially.

MALE. *Head*: brownish black, scattered white scales on vertex; labial palpi reaching slightly beyond antennal bases, prominent white streak on front of II. *Thorax*: brownish black with metallic scales green to blue-green; patagia with large dorsal and lateral white spots connected along fore margin to form collar; fore margin of tegulae broadly metallic green, lower mesal fringe with white scales showing from beneath; pectus brownish black, propleuron with conspicuous white patch; meso- and metapleura white-spotted; mesal area between midcoxae white; coxae and anterior margin of femora in forelegs prominently white; distal three segments of hind tarsi white. *Forewing*: outer half brownish black, median fascia metallic blue, iridescent scales extend narrowly basad along Sc and inner margin; oblique black streak in inner basal area with short metallic green streak above; underside with basal half metallic blue, except light brown in wing overlap. *Hindwing*: brownish black, iridescence absent except for scattered metallic scales below cell; area of wing overlap light brown; underside entirely metallic blue-green, except brownish black at tips. *Abdomen*: basal tergite brownish black with small metallic spot at center of caudal margin; remaining tergites golden-green, with two faint dark longitudinal striae subdorsally; basal and apical pleurites with distinct white spots; intervening pleurites with occasional white scales or small points; basal sternite with two subventral white spots and white scales at center; remaining six sternites brownish black, with series of white spots medially diminishing in size from base to apex; apical sternite with tiny white points lateral to those of medial series; lateral margins iridescent green. *Genitalia*: as in Figs. 71-74 (drawn from RED prep. 39221, USNM; 2n); valvae symmetrical, dorsal process (clasper) acuminate at tip with shorter thorn-like process on inner margin dorsocephalad; ventral process moderately hirsute, incurved beyond middle; dorsum of uncus with lateral margins extended as horizontal flanges from base; juxta unusually small, a thin sclerotized rod not reaching base of ventral process of valve; aedeagus with two spines dorsodistal to attachment of diaphragma; membrane sclerotized around point of attachment.

FEMALE. Differs from the male by: *Head*: without white scales on vertex; labial palpi without white streak on front of II. *Thorax*: iridescence blue, not green; dorsal and lateral white spots of patagia smaller and not connected along fore margin; pectus less white; white scaling absent from femora of forelegs and reduced to inner mesal margin and to small paired spots on proximal end of forecoxae dorsad and laterad. *Forewing*: more broadly suffused with metallic scales; medial fascia not defined, but forming part of generalized suffusion that may reach termen except at apex; oblique, black basal streak always present with short metallic streak above. *Hindwing*: small patch of blue scales between fork of cubitals at termen; underside with costal area and upper part of cell metallic green, metallic blue at fork of Cu_{1+2} and below A_3. *Abdomen*: iridescence on tergites appears dark green; distinct white spots in pleura of segments II, VI, and VII, rather than on II and VIII as in males; venter with only five white spots rather than six. *Genitalia*: as in Figs. 158-159 (drawn from RED prep. 39235, BMNH; 5n); sterigma with prominent pair of sclerotized pockets along cephalic margin of sternite VII; small cuneiform sclerite between pockets; segment VII more heavily sclerotized than anterior segments, with sternite V-shaped and symmetrical, caudal margin straight, emarginate mesad; small blind bursa on left dorsocephalad of ductus bursae; inner margin of lamella antevaginalis at ostium curved, with deep notch to left of center; lamella postvaginalis with distinctive longitudinal fold.

VARIATION. Length of forewing: males, 14.2 mm-16.5 mm; females, 15.3-17.0 mm. *In males*: white scales on vertex often absent; forecoxae occasionally brownish black distally rather than entirely white; metallic scales of median fascia of forewing green or blue-green rather than blue; upperside of hindwing sometimes devoid of metallic scales; iridescence on wing underside differing in color between forewing and hindwing, either blue, blue-green, or green; underside of hindwing sometimes devoid of metallic scales; longitudinal striae on abdomen may be distinct rather than faint. *In females*: metallic streak below cell of forewing usually a contrasting green but occasionally blue or green and not easily distinguished from surrounding scales; iridescence on forewing usually extending beyond tornus, but occasionally pattern resembling male (cf. males and females of *adonis*).

It is difficult to generalize about the variation in this species because there are always exceptions, depending on the populations examined. Fleming (1957) inaccurately observed that the white spot on the propleuron is absent in females. The spot is often reduced in females, and in specimens from Trinidad it is sometimes present only as a few scales. Females from mainland populations show a distinct spot, sometimes reduced compared to that in males.

TYPE DATA. ! *aurifera* Hampson: female holotype, La Merced, Perú, 2500' (Watkins), BMNH. ! *spinivalva* Fleming: male holotype, Trinidad, Simla, Arima Valley, 27 - III, N.Y. Zool. Soc. cat. no. 5716, AMNH.

BIOLOGY. No information.

GEOGRAPHICAL DISTRIBUTION. Trinidad, northern and western Venezuela, and eastern regions of Colombia, Ecuador, and Perú. A single record exists for southeastern Brazil.

FLIGHT PERIOD. Probably active throughout the year. There are adult collection records for every month except April, October, and November.

REMARKS. There are no records for males of *M. aurifera* from Perú, even though the type-locality is La Merced. This absence was perhaps responsible for Fleming's failure to recognize *aurifera* in Trinidad when he described a male as *spinivalva*. The discovery of the *spinivalva* synonymy came about when I realized Fleming's type series was mixed, and that among the paratypes were specimens with a 7th abdominal sternite appearing identical to that for *aurifera*. The paired intersegmental pockets and general shape of the sternite are unmistakable. An *in copulo* pair (N.Y. Zool. Soc. cat. no. 5718) enabled me to associate the sexes and verify the synonymy. The allotype of *spinivalva* appears to be a female of *lades*. Other specimens in this series (nos. 5728, 5729, and 5734) are either *thyra* or *thyridia*, but genitalia dissections are necessary for verification. Both species are recorded from Trinidad.

SPECIMENS EXAMINED (53): 26 males, 27 females. **COLOMBIA**: *Cundinamarca*: Medina, east Colombia, 500 m, VM. *Meta*: Río Guayuriba, SMM. **VENEZUELA**: *Aragua*: Rancho Grande, 1100 m, UCV. *Carabobo*: San Esteban, Las Quiguas, CM, USNM. *Mérida*: Mérida, BMNH. *Miranda*: Parque Nacional Guatopo, La Macanilla, 500 m, UCV; Parque Nacional Guatopo [El Lucero] 24 km N. Altagracia de Orituco, 640 m, UCV. *Yaracuy*: Aroa, UCV. **TRINIDAD**: *St. George East*: Arima Valley, USNM. *St. George West*: Ariapite Valley, BMNH, USNM; Hololo Mt. Road, St. Anns, CM; Maracas Valley, 150 ft, BMNH; Mt. Tucuche, 2-3000 ft, BMNH; St. Anns Valley, BMNH. *Caroni*: Tabaquite, Nariev District, BMNH. "St. George's" [Trinidad?], BMNH. **SURINAM**: "Surinam," BMNH. **ECUADOR**: *Pastaza*: Sarayacu, BMNH. **PERU**: *Junín*: La Merced, BMNH; Colonia Perené, El Campamento, CU. *San Martín*: Juanjuí, upper Amazons, BMNH. **BRAZIL**: *Goiás*: Leopoldo de Bulhões, BMNH.

Macrocneme durcata Dietz, new species

(Figs. 58-61, 156-157, 222; Map 20)

The bright blue iridescence broadly suffused over the wings makes this one of the most beautiful species in the genus. It is common in collections, but has been misidentified as *coerulescens* Dognin or *cyanea* Butler. In males the double-pronged juxta is unique. In females the prominent ventral pockets from the encircling pleura of segment VII are diagnostic. Similar pockets are found elsewhere only in females of *aurifera* Hampson. *M. coerulescens*, although sympatric with *durcata* in northern South America, is smaller and seldom as extensively blue on the wings. *M. cyanea* is restricted to southern Brazil.

MALE. *Head*: brownish black; labial palpi reaching base of antennae. *Thorax*: brownish black with iridescence green to blue-green; white points on patagia small, iridescent spots large and covering outer half of sclerite; tegulae with iridescent streak along anterior margin and partially mesad, white scales present on underside of tegulae but not along fringe; pectus without iridescence except for scattered scales in metacoxal grooves; brownish black hair-like scales overlie prominent white spots on meso- and metapleuron; legs brownish black; all tibiae with thin iridescent streaks; coxae and femoral grooves of mid and hind legs white; distal three segments of metatarsi white. *Forewing*: strikingly iridescent blue, brownish black at apex and termen; black streak from basal angle to distal third of anal fold; scattered black scales above streak and into cell; underside similarly iridescent blue, but less extensive than above, basal half iridescent, outer half brownish black; diffuse white spots basally; inner margin at wing overlap tan. *Hindwing*: bright iridescent blue; costal margin at wing overlap tan; anal area with brownish black hair-like scales overlying metallic scales; underside iridescent blue, apex brownish black. *Abdomen*: brownish black; iridescent scales restricted to faint mid-dorsal stria on tergum from posterior margin of A_1 to A_3; slightly broader striae along lateral margins of tergum and venter; basal sternite entirely white; small, mesal spots on succeeding 4 segments; white spot absent from pleurite II. *Genitalia*: as in Figs. 58-61 (drawn from paratype, RED prep. 39190, USNM; 4n); dorsal processes (claspers) of valvae symmetrical; ventral processes incurved at tips and extending beyond claspers; mesal sclerite between processes of valvae with prominent protuberance; dorsum of uncus with lateral margins appearing as auriculate flanges; juxta short, double-pronged, with points barely reaching base of ventral processes; dorsum of aedeagus strongly curved to right, with single horizontal spine on upper margin; vesica with three bursae, the smallest lying ventrad (not seen in illustration).

FEMALE. Similar to male, except general appearance darker due to reduction in white scaling. *Thorax*: white from underside of tegulae visible along fringe of mesal margin; pectus comparatively darker than in male, with white on meso- and meta pleura and coxae reduced to scattered scales or absent; white absent from femoral grooves of mid and hind legs and reduced to distal two segments on metatarsi. *Forewing*: distal two-thirds of wing bright iridescent blue, with color reaching termen and extending closer to apex than in male; base of wing more extensively black than

82   *University of California Publications in Entomology*

in male, with black scales as irregular streak from basal angle to costa; proximal edge of streak interrupted at three points by metallic blue-green scales. *Abdomen*: white spots of basal tergite smaller and ground color darker than male; venter with white restricted to three points on basal sternite and a medial row of small spots on next 5 sternites. *Genitalia*: as in Figs. 156-157 (drawn from RED prep. 39263, USNM; 3n); sternite VII triangular, two oblique carinae at center, sides subequal, apex continuing as small intersegmental sclerite in shape of right triangle, caudal margin with prominent medial emargination; encircling portion of pleura VII with two sclerotized pockets (see Fig. 156) from anterior margin; inner margin of lamella antevaginalis smooth, unmodified at ostium; ovate sclerite of lamella postvaginalis narrowed and decurved at ostial opening; ductus bursae short; corpus bursae and base of accessory bursa with concentric plicae; opposing, spinose, scallop-shaped signa at middle of corpus bursae; anterior apophyses as triangulate pockets on lateral edges of sternite VIII.

VARIATION. Length of forewing: male, 19.0-20.0 mm; female, 18.5-21.3 mm. Populations of *M. durcata* from Venezuela and Colombia have more extensively blue wing iridescence than those from the upper Amazon. In general, specimens from Perú (Sto. Domingo; Carabayá), Bolivia (Río Yapacaní; Río Chiripiri), and Brazil (Cuyabá) have more green in the iridescence than the same species from Venezuela. Occasional Venezuelan specimens (Aroa) have green on the inner margin of the forewing or a green to blue-green spot at the forewing base (El Guacharo). Rarely, a specimen is bright, shiny golden-green (one male, Km 125, El Dorado-Sta. Elena, Venezuela). *M. cupreipennis* Walker was named from a similarly golden-green female. Except for a slight difference in the iridescent hue, the *cupreipennis* holotype is remarkably like the green variant of *durcata*. Both specimens are atypical of the species they represent.

TYPE DATA. Holotype male: Aroa, Venezuela, (Schaus); USNM Type 73263, USNM. Allotype: 3 km N. El Guacharo, Monagas, Venezuela, 28/29-VIII-1975 (Dietz), USNM. Paratypes (144): 55 males; 89 females. **COLOMBIA**: *Antioquia*: Botero, 4000 ft, July (Hall), BMNH; Medellín, December (Apollinaire), AMNH, USNM. *Caldas*: Manizales (Gallego), USNM. *Cundinamarca*: Bogotá (Apollinaire), USNM; Cananche, January (deMathan), BMNH; Medina, 500 m (Fassl), USNM. *Magdalena*: Don Amo, 2000 ft, July (Smith), BMNH; Onaca, 2500 ft, August (Smith), BMNH; Valparaiso, 4500 ft, May (Smith), BMNH. *Norte de Santander*: Bella Vista, 2300 ft, January (Cawse-Morgan), AMNH. Cúcuta, [sic] "Venezuela", BMNH. *Valle*: Yumbo (Fassl), USNM. **VENEZUELA**: *Aragua*: La Isleta, Choroní, 200 m, April, May, July, November (Romero, Murtaugh, et al.), UCV; El Limon [Maracay], February (Fernández), UCV; Ocumare de la Costa, sea level, July (Gelbez, Olivo), UCV; Pozo del Diablo, El Limon [Maracay], November (Fernández), UCV; Rancho Grande, 1100 m, February, March, April, July, December (Dietz, Poole, Irwin, Kern, et al.), CAS, RML, UCV, USNM. *Barinas*: Reserva Forestal Ticoporo, 230 m, May (Gelbez, Salcedo), UCV. *Bolívar*: El Dorado-Sta. Elena, Km 125, 1100 m (Rosales, Gelbez), UCV. *Carabobo*: Las Quiguas nr. San Esteban (Klages), BMNH, CM. *Distrito Federal*: Caracas, BMNH. *Mérida*: Mérida (Briceño, et al.), BMNH, PM, USNM. *Miranda*: Guatopo, 400 m, August (Fernández, Rosales), UCV; Parque Nacional Guatopo, Agua Blanca, 500 m, May (Dietz, Salcedo), UCV; Parque

Nacional Guatopo [El Lucero] 24 km N. Altagracia de Orituco, 640 m, May (Dietz, Salcedo), UCV; Parque Nacional Guatopo, La Macanilla, 500 m, May (Dietz, Salcedo), UCV. *Zúlia*: El Tucuco, 420 m, May (Salcedo, Rosales, Ramirez), UCV. "Venezuela", (Joicey), BMNH. **ECUADOR**: *Zamora-Chinchipe*: Zamora, 3-4,000 ft (Baron), BMNH. *Imbabura*: Intaj [=Cordillera Intag], (Buckley), BMNH. *Tungurahua*: Rosario [?], Sta. Inéz [hacienda E. of Baños on Río Pastaza], 1250 m (Haensch), BMNH. **PERU**: *Junín*: Chanchamayo, 2100-7500 ft (Schunche), BMNH; Valle Chanchamayo, 800 m, August (Weyrauch), IML; vicinity of Sani Beni (=Río Sanibeni], 840 m, August (Woytkowski), CM. *Loreto*: Pumayacu (Johnson), USNM. *Puno*: Santo Domingo, Carabaya, 6000 ft, wet season, November-December (Ockenden, et al.), BMNH, USNM. **BOLIVIA**: *Cochabamba*: Yungas del Espiritu Santo, (Germain), BMNH; Chapare - [Upper region of] Río Chipiriri, 400 m, November (Förster), SMM. *La Paz*: Río Zongo [=R. Songo], 750 m (Fassl), PM, VM; Yungas de Coroico, 1800 m (Garlepp), BMNH; Yungas de la Paz, 1000 m (Rolle, Garlepp), BMNH, USNM. *Sta. Cruz*: Provencia del Sara, 450 m, December (Steinbach), BMNH; Río Yapacaní, 600 m (Steinbach), CM. **BRAZIL**: *Mato Grosso*: Buriti, 30 miles NE Cuyabá, 2250 ft, July (Collenette), BMNH; chapada near Cuyabá (Smith), CM. Non-paratypic material: The following localities are based on single females which I can only tentatively identify as *durcata* and therefore do not consider paratypes: BRAZIL: *Rio Grande do Sul*: Porto Alegre (Rolle), PM; Rio Grande do Sul, PM. "Amazon," PM.

BIOLOGY. Two females taken at light in January at Rancho Grande, Venezuela, oviposited in polyethylene bags, laying clusters of 15 and 28 eggs, respectively. Larvae reared on the modified Shorey diet failed to reach maturity. Following are descriptions for the available stages.

Egg: Pale yellow, semi-spherical, shiny, smooth, 1.23 mm wide, chorion transparent with tiny hexagonal reticulations over surface; eggs deposited singly or in small clusters; eggs not in contact with each other; eclosion occurred 6-7 days after oviposition; larvae consumed chorion upon emergence.

Larva: **Instar 1**: length 7.8 mm; head capsule light brown, width 0.54 mm; body integument white; setae plumose, arising from sclerotized chalazae; setae D and SD black, remaining setae colorless; seta D1 shorter than D2 or SD on abdominal segments; secondary setae absent. **Instar 2**: length 7.8 mm; head capsule dark brown to black, width 0.65 mm; body integument grayish white, surface spinulate; two rows of prominent verrucae formed by D+SD and L of thorax and D2 and SD of abdomen; D1 of abdomen tiny, equidistant, not contiguous on A1; single white seta from verruca L on T3 and SD on A8. **Instar (4?)**: moult not completed; head capsule dark brown to black, width 1.04 mm; instar differs from preceding larva by having bases of verrucae (especially D and SD) iridescent blue and an additional white seta from verruca D+SD on T3. **Final Instar**: not available.

Pupa: Not known.

GEOGRAPHICAL DISTRIBUTION. Venezuela, northern and central Colombia, upper Amazonian regions in Ecuador, Perú, Bolivia, and Brazil.

FLIGHT PERIOD. Probably flies throughout the year. Collection records are available for every month of the year except September and October.

ETYMOLOGY. The specific epithet is a Schaus manuscript name taken from a hand-labeled specimen in the collection of the USNM. The name *durca* was used by Schaus for a species of *Cosmosoma* in 1896 with no indication of its origin; *durcata* appears to be a derivative.

REMARKS. Although *M. aurifera* is smaller than *durcata*, its genitalia suggest a phyletic affinity. In males of *durcata* the juxta is reduced to a short, two-pronged structure, while in *aurifera* it almost disappears as a tiny rod-like band. All other *Macrocneme* have a prominent, strong juxta. In females of both species the sterigma is similar in having paired intersegmental pockets between sternites VI and VII.

Macrocneme bodoquero Dietz, new species

(Figs. 84-87, 160-161, 223; Map 21)

This is a comparatively small species for *Macrocneme*, similar in size and in its Amazonian distribution to *M. zongonata* but distinguished by the absence of a basal green dash on the forewing and by the possession of black hind tarsi. An affinity with *M. lades* is suggested by the structural resemblance of the male genitalia except for the aedeagal spines (cf. Figs. 50 and 87). The absence of white on the hind tarsi, and a more suffuse wing iridescence distinguish *bodoquero* when sympatric with *lades*.

MALE. *Head*: brownish black, frons mostly white; labial palpi almost reach base of antennae. *Thorax* (including legs): brownish black, disc iridescent green with non-iridescent hair-like scales from sclerite borders; white and metallic spots of patagia comparatively small; metallic spots on tegulae small, white absent from underside; pectus brownish black with white spot on each segment, the prothoracic spot smaller than either meso- or metathoracic spot; coxal grooves, fore and mid tibiae scaled with iridescent blue, tips of hind tarsi black. *Forewing*: brownish black, basal half iridescent blue with scales reaching beyond cell, veins dark, oblique black streak from basal angle not reaching bottom of cell; underside similar to upperside, except iridescence not extending beyond cell and inner margin tan where wings overlap, retinaculum brownish black. *Hindwing*: brownish black, discal area iridescent blue; underside iridescent blue except at apex and in basal area below cell, veins dark. *Abdomen*: brownish black, basal white spots on dorsum small, iridescence restricted to three longitudinal green stripes, one mid-dorsal and two lateral, venter with slight iridescence along lateral margins, basal sternite mostly white, attenuating series of small white spots mesad. *Genitalia*: as in Figs. 84-87 (drawn from RED prep. 39220, USNM; 2n); dorsal processes (claspers) of valvae slightly asymmetrical, with apex of left arm wider and more sharply pointed than corresponding apex on right arm; mesal sclerite between clasper arm and ventral process prominently knobbed; uncus skewed slightly to left when viewed dorsally, dorsum slightly rounded, lateral margins extended as asymmetrical flanges, with left side broader than right; juxta always extends to base of ventral processes of valvae, lateral margins indented at middle, apical margin incised at right, tip edged with row of small spines (6-7); dorsum of aedeagus with two spines, the left stout, long, and surface spinulate, the right smooth, stubby, one-third length of left spine.

FEMALE. Essentially as described for male except: iridescence of forewing blue-green, especially basad; hindwing blue; basal abdominal sternite with three spots rather than all white; sternites VI and VII with white spots laterally. *Genitalia*: as in Figs. 160-161 (drawn from RED prep. 39262, BMNH; 2n); sternite VII of sterigma shield-shaped, with two small, concave sclerites intersegmentally, one unattached, the other extending from edge of encircling pleurite; thickened plicae ventrad to spiracles on segments V and VI; lamella antevaginalis half as wide as overlying VII, inner margin as illustrated; lamella postvaginalis a narrowly ovoid sclerite decurved at ostium; ostium bursae and ductus bursae partially sclerotized where membranes join; thickened plicae in dorsal wall of ductus bursae; accessory bursa comparatively short-stalked; bursa copulatrix and signa as described for genus.

VARIATION. Length of forewing: male, 17.0 mm; female, 16.0 mm. White scales sometimes present on underside of tegulae.

TYPE DATA. Holotype male: Colombia, Caquetá, Morelia, Río Bodoquero, 430 m, 19-20/I/69, Duckworth and Dietz, USNM. Allotype: Perú, Pebas, Amazones, fin X^{bre} & 1er Tr. 1880 [= December-March, 1880], M. deMathan, ex Oberthür Coll., Brit. Mus. 1927-3, BMNH. Paratypes (24): 19 males, 5 females. **COLOMBIA**: *Caquetá*: Morelia, Río Bodoquero, 430 m, January (Duckworth & Dietz), USNM. **PERU**: *Loreto*: Pebas, December-March (deMathan), BMNH. **BRAZIL**: *Amazonas*: Hyutanahan [=Huitanaã], Río Purús, January-March (Klages), CM; Manaos [= Manaus], PM; São Paulo de Olivença, December (Wucherpfenning), BMNH; Teffe [Tefé], September, November (Fassl, deMathan), BMNH, USNM. *Pará*: Pará [Belém], (Moss), BMNH. *São Paulo*: [Serra da] Cantareira, April (Spitz), BMNH. No locality: one male, ex Holland Collection, CM.

BIOLOGY. No information.

GEOGRAPHICAL DISTRIBUTION. Amazon basin of Colombia, Perú, and Brazil. One disjunct record from Cantareira, Brazil, is based on a single male.

FLIGHT PERIOD. Adults have been collected from November through April and once in September. They probably fly year-round except in months of heavy rainfall.

ETYMOLOGY. The specific epithet is taken from the type-locality where the Spanish *bodoquero* (=blowgun hunter) might have been what an early entomologist encountered when collecting in these foothills of the Eastern Cordillera of Colombia.

REMARKS. Attention is drawn to the diminutive size of the white markings in this species, compared to most other *Macrocneme,* and to the contrasting blue iridescent wings against a green thorax and abdomen.

Macrocneme imbellis Dietz, new species

(Figs. 88-91, 224; Map 22)

This is a small Amazonian species with similarities in the genitalia that suggest
an alliance with *M. melanopeza*. It is described from a unique male with a prominent
white underside. Its size and phenotype resemble the "intacta" form of *thyra*, but its
palpi are porrect and the juxta is long and spinulose along the entire apical margin.
    MALE. Length of forewing: 15.5 mm. *Head*: brownish black, occiput with two
large metallic spots, frons mostly white with spots below antennae large, contiguous;
labial palpi short, porrect, basal segment (I) entirely white, segment II white on outer
surface, segment III brownish black. *Thorax*: brownish black, iridescent markings of
disc, patagia, and tegulae green; metallic spots of patagia prominent; underside of
tegulae white with scales visible on mesal fringe; pectus strongly white, except
brownish black beneath wings, propleuron prominently white-spotted; coxae white,
front femora [missing in type] presumably like mid femora cream above, white below;
all tibiae and hind femora brownish black above, white below; hind tarsi
white-tipped, with white strongest on inside margin. *Forewing*: brownish black, small
metallic green dash at base, thin iridescent blue streaks medially with heaviest scaling
on inner margin; black scales from basal angle not apparent; underside iridescent
light blue to end of cell, apex brownish black, Cu_2 white-scaled, broad white streak
through anal fold to tornus, inner margin tawny, retinaculum white. *Hindwing*:
brownish black, iridescent scales absent, prominent white patch at base; underside
white basally, iridescent light blue distally to apex. *Abdomen*: dorsum and pleural
region strongly iridescent green, venter entirely white, basal tergite brownish black
with usual four white spots, single white spot in pleura basad. *Genitalia*: as in Figs.
88-91 (drawn from holotype, RED prep. 278, BMNH; 1n); dorsal processes (claspers)
of valvae nearly symmetrical, left arm slightly wider than right, tips curved dorsad,
inner surfaces concave; thin, flat scales interspersed among setae on ventral
processes; uncus only slightly skewed left when viewed dorsally, dorsum lightly
rounded, lateral margins narrowly flanged; juxta long, somewhat spatulate, extending
well above base of ventral processes, apex incised at right, remaining margin
spinulose along inner surface; dorsum of aedeagus with two short spines directed
dextrad, left tip sharply acuminate, right tip round.
    TYPE DATA. Male holotype: Iquitos, Perú, upper Amazon, March 1932 (G.
Klug), BMNH.
    BIOLOGY. No information.
    GEOGRAPHICAL DISTRIBUTION. Known only from the type locality, Iquitos,
Perú.
    FLIGHT PERIOD. Insufficient information.
    ETYMOLOGY. The specific epithet is taken from the Latin adjective *imbellis*,
meaning unwarlike, peaceable.
    REMARKS. Only the holotype was examined. The ventral processes of the valvae
have long triangulate scales interspersed among the setae. These scales are also
found in *M. melanopeza, habroceladon, ferrea*, and *cyanea*. Whether the common

possession of these scales is of phyletic significance is unknown, but it has only been seen in two other species, *ferrea* and *cyanea*, and these are restricted to southern Brazil.

Macrocneme melanopeza Dietz, new species

(Figs. 96-99, 235; Map 22)

This ia a medium-sized Andean species described on the basis of two males from Pan de Azúcar, Perú. It appears allied to *Macrocneme imbellis* by similarities in the genitalia, but is distinguished by bright blue wings, black hind tarsi, and the absence of heavy white scaling on the underside of the thorax and abdomen. As in *imbellis*, the ventral processes of the valvae have scales interspersed among the setae and the juxta is spined along the entire apical margin. The juxta differs in *melanopeza* by not being incised on the right margin. Also, the shape of the claspers and of the uncus dorsum separate the two species.

MALE. Length of forewing: 16.5 mm. *Head*: brownish black, metallic blue on vertex and on spots on occiput; labial palpi upturned, not reaching base of antennae. *Thorax* (including legs): brownish black, iridescent scales blue; white spots of patagia small, metallic blue restricted to few scales adjacent to lateral white spots; tegulae without metallic scales, white absent from underside; pectus brownish black with white spots above coxae, especially prominent on proleuron; forelegs missing in type specimen, but paratype male shows forelegs to have usual white and metallic markings described for genus; coxal grooves lined with blue; tibiae streaked with blue, hind tarsi black. *Forewing*: largely suffused with iridescent steel blue, brownish black at apex, black streak from basal angle; underside metallic blue basally, brownish black distally; retinaculum brownish black with tip off-white. *Hindwing*: discal area to outer margin and above anal fold metallic blue, brownish black otherwise; underside similarly blue basally, brownish black apically. *Abdomen*: dark green with a thin dorsal stria and lateral margins iridescent green to blue-green, iridescence becomes obsolescent at mid-abdomen; venter with iridescence along lateral margins; single white spot in pleura; white spots of mid-ventral series large basally, obsolescent apically; sternite I with large lateral white spots. *Genitalia*: as in Figs. 96-99 (drawn from holotype, RED prep. 39241, LACM; 2n); dorsal processes (claspers) of valvae slightly asymmetrical, left arm somewhat longer and less curved than right arm; thin flat scales interspersed among setae on ventral process; uncus not skewed to left when viewed dorsally, dorsum strongly convex, lateral margins forming vertical flanges; juxta large, spatulate, extending well above base of ventral processes, apical margin spined, with tips pointing to right, right margin extended into flap with marginal spines pointing in opposite direction (see Fig. 98); dorsum of aedeagus with two small spines directed dextrad.

FEMALE. Unknown.

VARIATION. In single paratype, iridescence more heavily scaled, making wings appear brighter, shinier blue than described for the type; patagia with strong metallic blue spots; tegulae blue-streaked across foremargin and partially mesad.

TYPE DATA. Male holotype: Pan de Azúcar, Dept. Pasco, Perú, July 8, 1961 (F.S. Truxal), LACM. Only one other specimen was examined, a paratype male with the same data as the type, except date is July 6 instead of July 8.

BIOLOGY. No information.

GEOGRAPHICAL DISTRIBUTION. Known only from the type locality in Perú.

FLIGHT PERIOD. Insufficient information to determine.

ETYMOLOGY. The specific epithet, describing the hind tarsi of this species, is derived from the Greek *melanos* = black and *peza* = foot.

REMARKS. The blue and extent of iridescence on the wings are similar to that found in the blue variant of *coerulescens q.v.* These latter specimens differ in that they are larger and the hind tarsi are white-tipped.

Macrocneme orichalcea Dietz, new species

(Figs. 79-83, 164-165, 226; Map 22)

This is primarily an Amazonian species, often iridescent blue, readily recognized by two short black streaks at the base of the forewing, the lowermost always the more prominent. The species appears allied to *Macrocneme zongonata* by its distribution and to *aurifera* by genitalic resemblance. *Orichalcea* lacks the white spot on the occiput and the basal green streak of *zongonata*. Also, the hind tarsi are entirely black rather than white-tipped. Whereas *aurifera* has both clasper arms spined on the inner margin, only the left arm is spined in *orichalcea*.

MALE. *Head*: brownish black, lower frons white-scaled, vertex lightly metallic, occiput double-spotted, blue-green; labial palpi not reaching base of antennae, segments II and III smooth, tip of III slightly porrect. *Thorax* (including legs): brownish black, disc iridescent green; patagia with blue-green spots adjacent to lateral white markings; tegulae with short metallic streak anteromesad, white absent from underside; pectus rough-scaled, with iridescence restricted to coxal grooves, white spot of propleuron faint; forecoxae white on inner mesal surface and possessing two white spots anterolaterad, metallic scales scarce on forecoxae but forming streaks on fore and mid tarsi, hind tarsi black-tipped. *Forewing*: mostly iridescent blue or slightly greenish blue basad depending on angle of light, apical margin thinly non-metallic; prominent basal black streak between anal fold and A_2, thinner black streak above separated by line of iridescence along cell. Underside: base to end of cell iridescent blue except gray brown where wings overlap, outer half and retinaculum brownish black. *Hindwing*: brownish black with outer discal and limbal areas iridescent blue. Underside: mostly brownish black, iridescent blue along costa, in upper half of cell, and in broken mesal band. *Abdomen*: dark brownish black, with mid dorsum and lateral margins longitudinally striped iridescent green; venter similar, except small white spots in mesal series and additional white spot basad in pleura.

Genitalia: as in Figs. 79-83 (drawn from RED prep. 39222, UCV; 2n); dorsal processes (claspers) of valvae asymmetrical, tips acuminate, right arm more curved than left, lacking thorn-like process found on inner dorsocephalic margin of left arm; mesal sclerite of valve prominent, knobbed; dorsum of uncus narrow, lateral margins as vertical flanges; juxta extending well above base of ventral processes, apex broadly incised at right with lower corner acuminate, mesal margin membranous, tip margined with uneven spines, scattered spines basad on inner sclerotized fold; aedeagus with two spines, both perpendicular to dorsum.

FEMALE. Essentially the same as the male, except: vertex of head more strongly metallic; white on inner mesal surface of forecoxae not visible; color of forewing iridescence noticeably greener; blue iridescence on hindwing slightly reduced. *Genitalia*: as in Figs. 164-165 (drawn from RED prep. 266, BMNH; 1n); sterigma with sternite VII roughly scutiform, apex truncated, skewing right, caudal margin uneven, slightly notched at center; encircling pleurites of VII bullate, with thin sclerotized folds along anterior margins; intersegmental cuticula between VI and VII with various irregular sclerotizations (see Fig. 164); lamella antevaginalis as inner fold of sternite VII, margin unmodified at ostium; lamella postvaginalis roughly ovate, decurved slightly along ostial opening; thickened plicae in dorsal wall of ductus bursae at left; membranes of corpus bursae and accessory bursae in concentric plicae; signa with recumbent spines.

VARIATION. Length of forewing: males, 15-17 mm; females, 15.4-16.8 mm. Color of wing iridescence varies within a population, e.g., holotype male mostly blue, allotype female distinctly green (both specimens appear to have been reared from same lot); iridescence varies in extent of coverage on forewing, usually reaching apex, but may obsolesce before tip in some specimens; in male genitalia left clasper arm occasionally variable (Fig. 80), tip appearing truncated.

TYPE DATA. Holotype male: Pará, Brazil (A.M. Moss), Pupa no. 15, Rothschild Bequest, B.M. 1939-1, BMNH. Allotype: same as for male, except ex Pupa no. 71, BMNH. Paratypes (42): 21 males, 21 females. **COLOMBIA**: *Amazonas*: Loreto Yacú, April (Richter), SMM. *Putumayo*: Mocoa, 530 m, February (Hopp), PM. **VENEZUELA**: *Amazonas (Territorio Federal)*: Yavita, February (Lichy), UCV. *Bolívar*: El Dorado-Sta. Elena, Km 107, 520 m, August (Fernández, Rosales), UCV; El Hormiguero, Meseta de Nuria, 500 m, December (Expedición Instituto Zoología Agricola), UCV. **ECUADOR**: *Napo*: Apuya near Napo, January (Descimon), PM. **PERU**: *Loreto*: Río Marañón, BMNH. **BOLIVIA**: *Santa Cruz*: Provencia del Sara (Steinbach), CM. **BRAZIL**: *Amazonas*: São Paulo de Olivença, February, August, October (de Mathan, Wucherpfenning), BMNH; Igarapé Prêto, September (Waehner), BMNH. *Pará*: Oyapock, May, NMB; Pará (Bates, Moss), BMNH. No other data: "Amer. Mer." (Boisduval collection), BMNH. Non-paratypic material: 1 female, "Rio" (Hanson), BMNH.

BIOLOGY. No larvae available for study. From preserved exuvium of allotype female (Pupa 71, Pará, A.W. Moss), the dark brown head capsule of the final instar measured 1.93 mm wide and the verrucae remains were iridescent steel blue, much like other species in *Macrocneme* (cf. *coerulescens, thyra*).

Pupa (from same exuvium): length of shell 13.0 mm; medium brown with striped aposematic appearance of other *Macrocneme* pupae; appendages outlined in black and abdominal segments banded black along margins; pupa enclosed in a white, loosely-woven cocoon with black (and a few white) plumose setae interwoven over the surface.

GEOGRAPHICAL DISTRIBUTION. Eastern Venezuela and throughout the Amazon Basin, including tributaries in Brazil, Colombia, Perú, Ecuador, and Bolivia. The "Rio" locality is based on a single female specimen, indicating that if the identification is correct the species is rare in southeastern Brazil.

FLIGHT PERIOD. Probably flies throughout the year; collection records are available for every month of the year except March, June, July, and November.

ETYMOLOGY. The specific epithet is derived from the Latin *orichalcum*, meaning yellow-gold ore or the metal made from it, i.e., brass. The name was taken from a manuscript label on a female specimen from the Boisduval Collection in the British Museum.

REMARKS. The locality for the Boisduval specimen (above) is "Amer[ica] Mer[idionalis]," i.e., South America. To avoid the generality of this locality designation and to tie the name *orichalcea* to the more readily distinguished males, I have selected a holotype from Pará, Brazil. The allotype female is clearly associated.

Macrocneme zongonata Dietz, new species

(Figs. 92-95, 171-172, 227; Map 23)

This Amazonian species may be recognized by a white spot on the occiput and by the dark blue wings with a distinctive green dash encircled in black at the forewing base. By distribution and similarities in the male genitalia (pointed claspers, aedeagal spines), a relationship with *M. orichalcea* is possible. The genitalia in both sexes are diagnostic, especially the configuration of sternite VII in the females and the non-flanged uncus and rectangular juxta without spines in the males.

MALE. Length of forewing:15.5 mm. *Head*: brownish black, white scales on occiput flanked posteriorly by small metallic spots; labial palpi not reaching base of antennae. *Thorax* (including legs): brownish black, metallic marking on disc, patagia, and tegulae bright green, iridescence on disc not obscured by overlying hairs; patagia with metallic scales scarce, restricted narrowly to anterior margin; tegulae with prominent metallic streak across shoulder, underside white with tips visible along mesal fringe; pectus with white spots above coxae, largest spot on metepimeron; inner margin of forecoxae with white scales scarce; tibiae with weak iridescent streaks, hind coxal grooves metallic-lined, hind tarsi white-tipped. *Forewing*: bright metallic green at base, surrounded by black to medial area, outer half deep blue, apex and fringe brownish black; underside brownish black, with blue streaks along costa and lower edge of cell, white suffusion at base, retinaculum white. *Hindwing*: brownish black with faint scattering of deep blue beyond cell, base with two white spots; underside with prominent green to blue-green streak along costa and in upper half of cell, scattered blue scales at anal angle. *Abdomen*: brownish black, tergites II and III bright metallic green with iridescence extended narrowly along lateral margins of remaining segments, basal white spot in pleura, venter brownish black, lateral margins metallic green, white spots of mesal series diminish in size caudally. *Genitalia*: as in Figs. 92-95 (drawn from paratype, RED prep. 39211, BMNH; 2n); dorsal processes (claspers) of valvae nearly symmetrical, right arm slightly more curved than left arm, inner surfaces convex, tips narrow, acuminate; dorsum of uncus lightly convex, lateral margins round, not extended into flanges; juxta rectangular, apex extending slightly above base of ventral processes, apical margin extended to sharp points laterally; dorsum of aedeagus with two large spines, equal in length, sharp-pointed, tip of left spine slightly hooked to outside.

FEMALE. Length of forewing:16.3 mm. Essentially identical to male, except pleura of abdomen with four white spots rather than one; mesal series on venter with large spots that do not obsolesce caudally. *Genitalia*: as in Figs. 171-172 (drawn from allotype, RED prep. 281, USNM; 2n); sternite VII of sterigma u-shaped, symmetrical, apical margin not skewed, appearing truncated, broadly continuous with large intersegmental sclerite having two shallow indentations medially; medial protuberance of sternite VIII without usual thickened plicae below or laterally; inner fold of lamella antevaginalis strongly concave; interior margin even; sclerite of lamella postvaginalis roughly ovate, indented medially, not inserted at ostium; spines of signa comparatively long for genus.

TYPE DATA. Male holotype: Brazil, Amazonas, São Paulo de Olivença, November-December (Fassl), (Dognin Collection), USNM Type 73267, USNM. Allotype: same data as male (no type #), USNM. Paratypes (34): 26 males, 8 females. **PERU**: *Puno*: La Oroya, Rio Inambari, 3100 ft., Carabaya, October-December, wet season (Ockenden), BMNH; La Unión, Río Huacamayo, Carabaya, 2000 ft, December, wet season (Ockenden), BMNH. **BOLIVIA**: *Cochabamba*: Chapare, [Alto] Río Chipiriri, 400 m, November (Förster), SMM. *La Paz*: Río Songo [=R. Zongo], 750 m (Fassl), PM. *Santa Cruz*: Buenavista, 750 m (Steinbach), BMNH; Provencia del Sara, 450 m, November (Steinbach), CM. **BRAZIL**: *Amazonas*: Fonte Boa, July, September, October (Klages), BMNH; Humayta [=Humaitá], Rio Madeira, July-September (Hoffmanns), BMNH; Hyutanahan [=Huitanaã], Rio Purús, February (Klages), CM; Manacapurú, March (Klages), CM; Manaus, September, November (de Mathan), BMNH, CM; Miracema, Rio Purús, April (Klages), CM; Nova Olinda, Rio Purús, May (Klages), CM; Rio Manes [R. Maués], USNM; Rio Negro, iguapo (permanently flooded forest), 5 Km S. Rio Araras, April (Munroe), CNC; Rio Solimões, below R. Putumayo, September, CU; São Paulo de Olivença, November-December (Fassl), USNM; Teffé [Tefé], September, November (Fassl), BMNH; Tonantins, October (de Mathan), BMNH. *Pará*: Pará [=Belém], (Moss), BMNH; Taperinha nr. Santarém, September (Zerny), VM.

BIOLOGY. No information.

GEOGRAPHICAL DISTRIBUTION. Lower and upper regions of the Amazon Basin, Brazil, Perú, and Bolivia.

FLIGHT PERIOD. Appears to fly throughout the year. Adult collection records are available for all months but January and June.

ETYMOLOGY. The specific epithet is derived from the locality, Río Songo [=R. Zongo], a site often appearing in Lepidoptera taxonomy due to early collections by

Fassl. His efforts have assisted greatly in expanding our distributional knowledge of the South American fauna, including unique collections of four species of *Macrocneme* from Río Zongo.

Macrocneme bestia Dietz, new species

(Figs. 104-107, 187-188, 228; Map 23)

This species resembles *M. thyra* but is restricted in distribution to southeastern Brazil, where it forms part of a species complex with *cyanea, cupreipennis,* and *pelotas.* The males are distinguished by the left spine of the aedeagus hooked at the tip and turning sharply outward. The females are not readily distinguished, except by their association with males from identical localities.

MALE. Length of forewing: 16.8 mm. *Head*: brownish black, occiput with metallic blue-green spots; labial palpi not reaching base of antennae. *Thorax* (including pectus and legs): brownish black, iridescence blue-green; metallic spots of patagia large, at center of sclerite, lateral white spots small; tegulae with metallic streak across shoulder and mesocaudad, underside with few white scales, not visible from above; metallic scales absent from pectus, white restricted to small spot on propleuron and larger spots on meso- and metapleuron; inside margin of forecoxae white, outer surfaces of tibiae with thin blue streak, hind tarsi white-tipped. *Forewing*: brownish black, with basal half metallic green, veins dark, outer margin of iridescence reaching end of cell, short black streak from basal angle not touching cell; underside brownish black, with basal third metallic blue except where wings overlap, retinaculum white. *Hindwing*: brownish black, discal area with blue metallic scales lying mostly below cell, not extending to outer margin or into anal area, costal margin tan; underside mostly metallic blue except brownish black at apex and outer margin, veins dark. *Abdomen*: dark metallic green, basal tergite and venter brownish black, longitudinal stripes not visible, basal sternite entirely white, remaining segments with mesal series of white spots diminishing in size caudally. *Genitalia*: as in Figs. 104-107(drawn from RED prep. 39205, SMM; 2n); dorsal processes (claspers) of valvae slightly asymmetrical, left arm more curved and wider distally than right arm, inner margin of left arm weakly emarginate near tip, corresponding margin on right arm smooth; dorsum of uncus slightly convex, lateral margins asymmetrically flanged; apex of juxta squarely truncate, barely reaching ventral processes of valvae, corners extended to points, patch of spines in membrane at left small; dorsum of aedeagus with two spines, the right short and straight, the left distinctly longer and sharply turned outward at tip.

FEMALE. Length of forewing: 18.4 mm. Essentially identical to male except in few minor details: coxal grooves with metallic scales; basal abdominal sternite three-spotted rather than entirely white; additional pair of white spots lateral to mesal series on sternite VII. *Genitalia*: as in Figs. 187-188 (drawn from RED prep. 261, SMM; 2n); sternite VII of sterigma shield-shaped, with caudal margin broadly incised, slightly elevated at right, apex cephalad, broadly round, not skewed; small sclerotized pocket to right of mid-line in intersegmental membrane between VI and VII; fold of lamella antevaginalis strongly convex medially, inner margin extended as broad sulcus on right; long, spatulate sclerite forms lamella postvaginalis, bent at middle with base extending into ductus bursae; pockets serving as anterior apophyses on segment VIII prominent and well inflated; ductus bursae without thickened fold or pouch on dorsal wall; corpus bursae and stalk of accessory bursae with concentric plicae; signa formed by two opposing spiny patches with recumbent, unequal spines.

VARIATION. Length of forewing: males, 16.8-18.6 mm; females, 16.1-18.4 mm. Wing iridescence variable, blue or green; black streak from basal angle sometimes broad, touching bottom of cell; iridescence of abdominal dorsum either dull and entire or as thin lateral and mid-dorsal stripes of bright metallic scales.

TYPE DATA. Holotype male and allotype: Brazil, Rio de Janeiro, Imbariê, 50 m, 1 and 3-V-1956, leg. H. Ebert, SMM. Paratypes (26): 12 males, 14 females. **BRAZIL**: *Rio de Janeiro*: Imbariê, 50 m, January, April, May (Ebert), SMM; Rio de Janeiro, May (Ebert), SMM; Xerém, April (Ebert), SMM; Angra dos Reis, Fazenda Japuhyba, July, August, October (L. Travassos Fº.), USP; *São Paulo*: Itanhaém, December (Munroe), CNC; Santos, May, USP.

BIOLOGY. No information.

GEOGRAPHICAL DISTRIBUTION. Coastal southeastern Brazil.

ETYMOLOGY. The specific epithet is taken from the Latin *bestia,* meaning animal.

FLIGHT PERIOD. Collection records are scanty, but as indicated for the paratypes, adults are known to fly in the months of January, April, May, July, August, October, and December. Given the usual availability of other *Macrocneme* species throughout the year, these gaps for *bestia* will most likely fill in with further collecting.

Macrocneme cupreipennis Walker

(Figs. 116-119, 175-176, 229; Map 24)

Macrocneme cupreipennis Walker, 1856: 1632

This species occurs in two forms, one rare and entirely golden-green, for which the species was named, and another that is typical *Macrocneme,* with metallic scales restricted to the basal half of the forewing and to the discal area in the hindwing. It is a member of the species complex from southern Brazil and northern Argentina that includes *M. leucostigma, pelotas,* and *cyanea.* Its normal appearance so closely parallels these sympatric species that only an examination of the genitalia will separate it definitively. The asymmetric claspers and shape of the uncus are diagnostic in males, while the shape of sternite VII and the intersegmental pocket will identify females.

The following description is *cupreipennis sens. lat.,* based on its usual appearance, while Walker's *cupreipennis sens. st.* is an uncommon variant (see Variation).

MALE. *Head*: brownish black, small metallic green spots on occiput; palpi not reaching base of antennae. *Thorax* (including legs): brownish black, iridescence of disc green, obscured somewhat by overlying non-metallic scales; patagia with prominent metallic spots adjacent to lateral white spots; tegulae with strong green streak anteromesad, white-scaled beneath, visible along mesal fringe; pectus with prominent white spots on pleura above each coxa; coxal grooves of hind legs metallic, hind tarsi white-tipped. *Forewing*: brownish black, metallic green spot below cell basally, separated from medial blue-green band to end of cell by oblique black streak from basal angle; underside with basal blue iridescence not reaching end of cell, inner margin tan where wings overlap, white as prominent basal spot and as thin streak along A_2; retinaculum white. *Hindwing*: brownish black, discal area iridescent blue-green, white along $Sc+R$ and basally; underside iridescent green except at apex. *Abdomen*: dark green with iridescent scales basally and as thin longitudinal lines mid-dorsally and laterally, white spots of basal tergite and sternite well defined, smaller and fainter in mesal series of venter. *Genitalia*: as in Figs. 116-119 (drawn from RED prep. 39168, USNM; 6n); dorsal processes (claspers) of valvae asymmetrical, left arm club-like toward apex, right arm long, slender to tip, inner surface of both arms concave, especially at apices; uncus when viewed dorsally skews left, dorsum proximal to base with small vertical flange at center, lateral margins strongly asymmetrical, with left side a broader, more distal flange than right side; juxta long, spatulate, extending almost to apices of ventral processes of valvae, apex entire, margined partially on left with short spines; aedeagus with two spines on dorsum directed dextrad, smaller at left, with single denticle on inner base, right spine larger, with denticles above and below base.

FEMALE. Essentially identical to male, including similar color and pattern of iridescent scaling. *Genitalia*: as in Figs. 175-176 (drawn from RED prep. 271, BMNH; 3n); sternite VII of sterigma broadly u-shaped, apex cephalad, skewed slightly to left and continuous with large single intersegmental pocket to right of mid-line, lateral flaps asymmetrical, caudal margin uneven; irregular ovoid sclerite of lamella antevaginalis bent at ostial opening; lamella postvaginalis half as wide as overlying portion of sternite VII; dorsal wall of ductus bursae without a thickened fold or pocket; spiny patches of signa comparatively large for genus.

VARIATION. Length of forewing: males, 18.5-21.0 mm; females, 18.0-20.2 mm. Ventral processes of male valvae sometimes strongly curved outward; wing iridescence variable in color and pattern; metallic scales reflect as green, blue-green, or blue, or sometimes as combination; only rarely do both wings reflect as bright golden-green (cf. holotype female); in one forewing pattern (Leme; Hamburg Velho) iridescence extends over basal half of wing, with a short black streak separating a basal spot (often green) from a broader medial band (often blue); discal area of hindwing metallic-scaled and underside blue basally; in less typical pattern (ex Hansa Humboldt) iridescence suffuses evenly outward from base, becoming obsolescent before apex, basal angle interrupted by a tiny black spot, color green, underside blue; in unusual instance with scales golden-green, both wings entirely metallic, basal black spot absent; underside similar, except outer half of forewing non-metallic; variant of latter pattern (Juquiá, Fonte Tapir) with iridescence reflecting a simple green rather than a golden-green, base interrupted by black spot below cell.

TYPE DATA. ! *cupreipennis* Walker: female holotype, BMNH. Type locality unknown, presumably Brazil.

Walker did not give a type locality in his original description, but his specimen has subsequently been labeled "Brazil". This designation is probably correct. I have seen only nine specimens that reflect the Walker description of *cupreipennis sens. str.* All are labeled "Brazil", except one from "Rio de Janeiro". Also, the species' distribution is entirely Brazilian, except for several records from the northern frontier of Argentina.

BIOLOGY. No information.

GEOGRAPHICAL DISTRIBUTION. Southeastern Brazil and northern Argentina.

FLIGHT PERIOD. Not well-defined. Scanty collection records indicate adults are available in all months except January, March, May, July, and September. They probably fly throughout the year.

REMARKS. At Oxford's University Museum there are five specimens (2 males, 3 females, ex Miers) of Walker's "gilded green" *cupreipennis*. I used these examples to associate the sexes and in determining, through a comparison of genitalia, that Walker's specimen is an atypical variant of the species. The usual appearance of *cupreipennis* is like any other *Macrocneme*. The cause and frequency of the "gilded green" variant is unknown, but its scarcity in collections suggests that it is a unusual genetic recombination, much like the single green specimen of *M. durcata* that turned up in eastern Venezuela. The genitalia will always identify it.

SPECIMENS EXAMINED (44): 32 males, 12 females. **ARGENTINA**: *Entre Rios*: La Soledad, close to frontier of Uruguay, BMNH. *Misiones*: Campo Viera, SMM. **BRAZIL**: *Paraná*: Curitiba, Cajurú, USP; Curitiba, Fonte Ahu, USP; Fernandes Pinheiro, 2600', BMNH; "Parana" (Janson), BMNH; Ponta Grossa, MCZ, USP. *Rio de Janeiro*: "Rio Janeiro", USNM; Petrópolis, VM. *Rio Grande do Sul*: Hamburgo Velho, SMM; Porto Alegre, BMNH. *Santa Catarina*: Hansa Humboldt [= Corupá], BMNH, USNM; Jaraguá do Sul, BMNH; Mafra, 800 m, BMNH; Nova Teutônia, CM; Rio Vermelho, 830 m, BMNH; "St. Catherines", USNM. *São Paulo*: Juquiá, Fazenda Poco Grande, USP; Juquiá, Fonte Tapir, 400 m, USP; Leme, SMM; São Paulo, 750 m (Jones), BMNH; Ubatuba, BMNH. "Brazil" (Menestries; Miers; Natterer), OX, PM, VM.

Macrocneme cyanea (Butler)

(Figs. 108-111, 173-174, 230; Map 25)

Mastigocera cyanea Butler, 1876:372.
Drucea cyanea.- Kirby, 1892:130.
Macrocneme cyanea.- Hampson, 1898:317.

This species is primarily from southern Brazil. One female is tentatively identified from northern Argentina. The presence of *cupreipennis* among other *Macrocneme* may be suspected if the appearance of the wing iridescence appears to be somewhat dull and granular. The male genitalia are diagnostic, and like *M. pelotas* there is a spiny patch on the inner membrane of the juxta.

MALE. *Head*: brownish black, occiput with small metallic blue spots; labial palpi not reaching base of antennae. *Thorax* (including legs): brownish black, iridescent blue of disc somewhat obscured by overlying non-metallic, hairy scales; blue spots of patagia large, occupying entire area between dorsal and lateral white spots; anterior margins of tegulae metallic blue, with a streak extending partially mesad, underside white but not visible along mesal fringe; white spot on propleuron small; coxal grooves white; fore and mid femora tan above, white beneath (including hind femora); tibia streaked blue on outside, white inside; hind tarsi white-tipped. *Forewing*: brownish black to black, with basal half bright metallic blue to end of cell, oblique black streak from basal angle to cell; underside blue to end of cell, white irrorations basally; retinaculum white. *Hindwing*: brownish black to black, discal area bright metallic blue; underside entirely metallic blue, two white spots basally. *Abdomen*: dorsum dull iridescent green, with shiny blue-green at base and as thin mid-dorsal stripe; pleura with single basal white spot; basal two sternites mostly white, remaining sternites thinly white from anterior margins. *Genitalia*: as in Figs. 108-111 (drawn from RED prep. 39238, USNM; 2n); dorsal processes (claspers) of valvae asymmetrical, left arm longer than right and curving dorsad at apex, tips expanded, concave on inner surface; ventral processes with flat, emarginate scales interspersed among usual setae; uncus skewed slightly to left when viewed from above, dorsum narrow with lateral margins as vertical flanges, unequal in length; juxta long, cuneiform, reaching beyond base of ventral processes, apex margined with short mesally-directed spines, small spinose patch on inner sclerotized membrane; dorsum of aedeagus with two short spines directed dextrad, subequal in length; diaphragma sclerotized at point of attachment.

FEMALE. Similar to male except: *Thorax*: tan scales absent from legs; white reduced to inner margins of forecoxae and proximal spots on all coxae; white spot on metepimeron; iridescence may suffuse over entire forewing except at apex or outer margin. *Genitalia*: as in Figs. 173-174 (drawn from RED prep. 273, BMNH; 2n); sternite VII of sterigma v-shaped, apex cephalad, skewed to left and continuous with broad intersegmental pocket at right between VI and VII; lateral margins forming asymmetrical flaps; caudal margin irregular but not emarginate; oval sclerite of lamella postvaginalis with margin decurved at ostium, not deeply inserted; ductus bursae without thickened plicae in dorsal wall; signa small, scallop-shaped, center spines smaller than peripheral spines.

VARIATION. Length of forewing: males, 18.0-20.0 mm; females, 17.5-20.5 mm. White on legs varies; inner margin of all tarsi sometimes strongly white; iridescence of abdominal dorsum blue or blue-green as well as pure green; white on abdominal venter in series of small mesal spots or as thin bands from the anterior margins of each segment.

TYPE DATA. ! *cyanea* Butler: female holotype, "Brazil", BMNH. Butler did not indicate whether his description was based on a unique specimen. Hampson (1898) indicated two females as "types" for *cyanea*, one labeled "Brazil", the other "Rio Janeiro". Both specimens have Butler handwritten labels on them, but only the

"Brazil" specimen bears the word "type." I consider this the holotype. The "Rio Janeiro" specimen came from the Walsingham collection and was not registered in the British Museum until 1879, as indicated by the figures 79-56 on the locality label (Watson, *in litt.*).

BIOLOGY. No information.

GEOGRAPHICAL DISTRIBUTION. Known from southern Brazil and northern Argentina.

FLIGHT PERIOD. All but one of the adult collection records fall between September and February, suggesting that the species is univoltine. The single female from Argentina collected sometime between mid-March and mid-June suggests that the activity period is longer than six months, but without a precise date it is not possible to determine whether the extension should include only one month or all four.

SPECIMENS EXAMINED (30): 16 males, 14 females. **ARGENTINA**: *Misiones*: [Coloñia] Eldorado, SMM. **BRAZIL**: *Minas Gerais*: Passa Quatro, Serra dos Cochos, 1400 m, BMNH; Viçosa, CU. *Pará*: "Para", SMM. *Rio de Janeiro*: Nova Friburgo, PM; Petrópolis, BMNH, SMM, USNM; Rio de Janeiro, BMNH, SMM, USNM. *Santa Catarina*: Hansa Humboldt [=Corupá], BMNH, USNM; "St. Catherines", USNM. *São Paulo*: Estaçao Biol. Boraceía, Salesópolis, 850 m, USP; São Paulo, BMNH. "Brazil", PM, ZIH. Country unknown: ex Musaeo Boisduval, one male (with mss. label "*Copaena auricalcea*"), and one female, BMNH.

Macrocneme ferrea Butler,   revised status

(Figs. 124-127, 179-180, 236; Map 26)

Macrocneme ferrea Butler, 1876:371.- Zerny, 1931b:241 (syn. of *M. leucostigma*).-
    Forbes, 1939:130 (syn. of *M. leucostigma*)
Macrocneme lades ab. *ferrea*.- Hampson, 1898:318.- Zerny, 1912:95

This little-known species is another in the complex from southeastern Brazil. It has been treated both as a form of *lades* and as a synonym of *leucostigma*. It is a valid species, presently known only from Espíritu Santo, and most closely allied to *M. cyanea, megacybe*, and possibly *cupreipennis*. A narrow uncus, asymmetrical claspers, and a spiney patch on the inner membrane of the juxta are traits shared with *cyanea*. Even though the wing iridescence is somewhat sparse in the specimens I have associated with *ferrea*, the golden-green color suggests an alliance with *cupreipennis (sen. str.)*. The sterigma in the *ferrea* type is unusual for the broad apex of sternite VII and for the small, well-formed intersegmental pocket. In both *cupreipennis* and *cyanea*, the apex is not as broad and the pocket is larger and shallower. The females of *megacybe* have a sterigma that closely resembles that found in *ferrea*. In fact, *megacybe* may be conspecific. (For further discussion, see *megacybe*.)

MALE. *Head*: brownish black, occiput with two tiny spots, metallic green; labial palpi short, not reaching base of antennae. *Thorax* (including legs): brownish black;

iridescence golden-green; disc metallic with thin, non-shiny scales overlying; patagia with large metallic spot adjacent to lateral white spot; tegulae with only short metallic streak from anterior margin, underside white with scales barely visible on mesal fringe; pectus hairy with usual white spots at proximal ends of fore and mid coxae and on pro- and metapleura; coxal grooves metallic-scaled; fore and mid tarsi metallic-streaked ventrally; hind tarsi white on distal segment only. *Forewing*: brownish black, basal half to slightly beyond cell suffused golden-green,veins dark; basal black spot visible only as scattered scales; underside similar, but metallic scales not extending beyond cell; retinaculum brownish black. *Hindwing*: brownish black, with large golden-green patch in discal and anal areas; underside entirely metallic green except at apex. *Abdomen*: brownish black, with iridescent green forming usual wide lateral and single thin mid-dorsal stripe; terminal segment (VIII) blue-green; basal pleurite with white irrorations only; mesal series diminishing in size apically. *Genitalia*: as in Figs. 124-127 (drawn from RED prep. 39240, USNM; 2n); dorsal processes (claspers) of valvae strongly asymmetrical, left arm longer and wider than short right arm, apex inflated; ventral processes with long flat, triangular scales intermixed with normal setae; uncus skews left when viewed above, dorsum narrow with lateral margins as subequal, convex flanges, the left distad to the right; juxta spatulate, reaching beyond base of ventral processes, apex briefly incised at right, remaining margin spined; oval patch of fine spines on inner membrane; dorsum of aedeagus with two spines, left long, surface spiculate, right distad, short with single denticle from upper surface; diaphragma sclerotized at point of attachment.

FEMALE. Similar to male, but iridescence green rather than golden-green. *Genitalia*: as in Figs. 179-180 (drawn from lectotype, RED prep. 39150, BMNH; 1n); sternite VII of sterigma broadly U-shaped, with apical margin skewed left and broadly truncate, caudal margins uneven; small intersegmental pocket invaginated at right between VI and VII, with thin sclerotized bands extending laterally; anterior margin of right pleurite reflexed slightly, forming shallow depression; reniform sclerite of lamella postvaginalis decurved at ostium, not deeply inserted; dorsal wall of ductus bursae with two thickened plicae; signa as described for genus.

VARIATION. Length of forewing: 16.3-19.6 mm. Wing iridescence green as well as golden-green; male genitalia may have additional spines on inner membrane of juxta opposite oval patch.

TYPE DATA. ! *ferrea* Butler: female lectotype, by Hampson (1898), Espíritu Santo, Brazil, (Stevens), BMNH.

Butler listed New Granada and Espíritu Santo as localities for his *ferrea* syntypes. The sex and number of syntypes was not specified. I recognize Hampson's entry of "Espíritu Santo, Brazil, 1 female type" as the lectotype designation for *ferrea*. I was unable to locate the syntype from New Granada in the British Museum. Most likely it was a different species since there is almost no overlapping between the Colombian and southern Brazilian species of *Macrocneme*.

BIOLOGY. No information.

GEOGRAPHICAL DISTRIBUTION. Known only from the state of Espíritu Santo, Brazil.

FLIGHT PERIOD. Of the three males from Santa Teresa, one was collected in May and the other two in September.

REMARKS. Butler originally distinguished *ferrea* from *leucostigma* by minor differences in the wing pattern. Such variation is common in *Macrocneme* and is seldom an indication of phyletic affinity. As common as *M. leucostigma* is in Brazil, I have not seen it from the state of Espíritu Santo. I have matched the sexes of *ferrea* by locality, which leaves open the question whether they are properly associated. I originally considered the three males from Santa Teresa to be a new species. No females were available. The *ferrea* type was a unique female, also from Espíritu Santo, but it matched no known species. Rather than propose a new name for the males, I have combined them with the *ferrea* female. A further discussion of the taxonomic possibilities related to this decision will be found under *megacybe*.

SPECIMENS EXAMINED (4): 3 males, 1 female (lectotype). **BRAZIL**: *Espíritu Santo*: Loureiro, 650 m, Santa Teresa, May (Nicolay), USNM; Santa Teresa, September (Munroe), CNC; "Espíritu Santo", (Stevens), BMNH.

Macrocneme leucostigma (Perty)

(Figs. 100-103, 181-182, 232; Map 27)

Glaucopis leucostigma Perty, 1833: 158, pl. 31, fig. 11
Macrocneme leucostigma.- Butler, 1876:371.- Kirby, 1892:128 .- Zerny, 1912:95 (synonym of *M. lades*).- Zerny, 1931b: 241.- Forbes, 1939: 130
Copaena maja.- Burmeister (*nec* F., 1787), Burmeister, 1878:387 [cited as *maja* Hübner, in error]
Copaena naja [sic].- [Misprint for *maja* Cramer] Zerny, 1912:95 (synonym of *M.lades*).- Draudt, 1916:103 (synonym of *M. lades*).- Zerny, 1931b:241 (synonym of *M. leucostigma*)
Macrocneme lades auct. (*nec* Cramer, 1776).- Hampson (in part), 1898:317.- Jorgensen, 1913: 74 (larva, pupa).- Draudt (in part), 1916:103
Macrocneme lades form *chionopus* Draudt, 1916:103.- Zerny, 1931b:241 (synonym of *M. leucostigma*)
Macrocneme lades chionopus Bryk, 1953:236 [NEW SYNONYMY]
Macrocneme iole.- Mabilde (*nec* Druce, 1884), 1896:155 (larva).- Costa Lima, 1950:237 (synonym of *M. chrysitis*) [misidentification]
Macrocneme chrysitis.- Costa Lima (*nec* Guérin, 1844), 1950:237; *ibid*., 1968:220 (foodplants) [misidentification]
Macrocneme chrysitis form *deceptans* Draudt, 1916:104.- Zerny, 1931b:241 (synonym of *M. chrysitis*) [NEW SYNONYMY]

This species is possibly the *maja* of Fabricius as yet not clearly defined. It is the most abundant and widespread species of *Macrocneme* in southern South America and is widely sympatric and similar in appearance to numerous congeners. It is easily identified by the two vertical flanges on the uncus dorsum in males and the V-shaped structure of the seventh sternite in females.

MALE. *Head*: brownish black, occiput with two small metallic spots laterally, labial palpi not reaching base of antennae. *Thorax* (including legs): brownish black, metallic scaling green; iridescence of disc obscured slightly by overlying hair-like scales; paired dorsal and lateral white spots on patagia small, metallic spots adjacent to lateral pair; anterior and mesal margin of tegulae prominently metallic-streaked, underside white, not visible from mesal fringe; propleuron and metepimeron of pectus with white spots; inner margin of forecoxae white, all coxae white-spotted proximally, hind tarsi white-tipped. *Forewing*: brownish black, basal half iridescent green to end of cell, oblique black streak from basal angle to costa, veins dark; underside similar, but iridescence more extensively blue than above and not reaching end of cell, black scales absent, inner margin of wing overlap non-iridescent, tan, with thin streak of white along vein A_2; retinaculum white. *Hindwing*: brownish black, discal area scaled with iridescent green, underside mostly iridescent green to blue-green except at apex. *Abdomen*: dark green with shiny metallic scales forming a thin mid-dorsal stripe and wider lateral bands especially basad; all pleural segments and mid venter spotted with series of small white spots. *Genitalia*: as in Figs. 100-103 (drawn from *in copulo* pair, RED prep. 39198, CM; 14n); dorsal arms (claspers) of valvae almost symmetrical, left arm slightly thicker, tip curving dorsad, right arm evenly slender to tip; uncus when viewed dorsally skews left, dorsum with two vertical flanges, right strong and constant, left weak and sometimes absent, lateral margins as vertical flanges, left side decidedly stronger than right; juxta long, reaching well beyond base of ventral processes, tip spined, incised at right, with margin extended into small thumb-like flaps; dorsum of aedeagus with two prominent spines, left strongly spinose on outer margin, right equal in length but slightly more slender, margins smooth; diaphragma sclerotized at attachment to aedeagus.

FEMALE. Essentially identical to male. Color and pattern of iridescence on wings and body similar in variation to males. White markings vary somewhat less, especially on coxae and abdominal venter, coverage restricted to defined spots rather than covering segment or structure. *Genitalia*: as in Figs. 181-182 (drawn from *in copulo* pair, RED prep. 39198, CM; 7n); sternite VII of sterigma v-shaped, apex cephalad, lateral margins as flaps, symmetrical, caudal margin slightly emarginate mesad; intersegmental membrane between VI and VII sclerotized and continuous with apex of sternite VII to form two broad pockets beneath encircling pleurites, right larger than left; vertical plicae (2) of sternite VIII to right of medial protuberance; bullae of anterior apophyses prominent; lamella antevaginalis with central portion concave forming broad groove to ostial opening; lamella postvaginalis a broad, comma-shaped sclerite, bent medially; ductus bursae with thickened fold in dorsal wall; membrane of bursa copulatrix and stalk of acessory bursa in concentric plicae; signa the usual opposing, scallop-shaped patches with recumbent spines.

VARIATION. Length of forewing: males, 16.9-20.9 mm; females, 17.2-20.3 mm. Iridescence of the wings either green, blue-green, or blue, and occasionally almost brassy. Iridescence usually consistent in color on both wings, but occasionally forewing green and hindwing blue; iridescence on underside usually more extensively blue than on upperside; black streak of forewing usually prominent, but occasionally reduced to a spot or absent entirely; white markings normally as well-defined spots, but occasionally spread to cover basal segments of abdominal venter, mid and hind coxae, or to form streaks on hind femora.

TYPE DATA. *leucostigma* Perty: Type lost. Type locality: Brazil, "montibus Prov. Minarum" [mountains of coastal province]. Form *chionopus* Draudt: Type not located. Type locality: Perú, (Coll. Schaus), presumably USNM [see below]. ! *chionopus* Bryk: female holotype, Roque, Perú, [26-III-1925], RNS. ! Form *deceptans* Draudt: female lectotype, **here designated**, Rio Grande do Sul, Brazil, BMNH.

Draudt's form name *chionopus* most likely refers to the same specimen that Hampson (1898, p. 318) designated "Ab. 2" of *M. lades*. I have been unable to locate this specimen, and cannot tell from Hampson's entry whether he meant the specimen was collected by Schaus but was part of the British Museum collection or whether he had seen a specimen on loan from Schaus at the USNM. Neither the British Museum nor the Paris Museum (a depository for some Draudt material) have the type. There is one female specimen that is possibly the type in the USNM labeled "Peru, " but without a collector or collection label or any notation to indicate that it was the specimen mentioned by Hampson.

BIOLOGY. Mabilde (1896) and Jorgensen (1913) have described a larva of *Macrocneme* from Brazil and Argentina, respectively. I have not seen this larva, but from their descriptions it appears they were describing the same species. I am considering it to be *M. leucostigma* Perty rather than *iole* Druce or *lades* Cramer as respectively named by Mabilde and Jorgensen, since *M. iole* is known only from Central America and *M. lades* is unknown from Argentina. Both authors comment on the abundance of the species, which makes me suspect that it was *leucostigma,* since this is the most commonly encountered *Macrocneme* in southern Brazil and northern Argentina. Following is Jorgensen's account of the larva (translated from the German):

> *Macrocneme lades* [sic]: The caterpillar [was] found in the forest near Bonpland on a low plant in September. It pupated at the beginning of October and the adult emerged a month later. The caterpillar is very similar to the arctiid larva: deep black with transverse rows of sky blue, button-shaped [verrucae] on all segments which bear long black setae; the segment borders have broad, flesh-red, transverse bands. Length 35 mm. The cocoon is large, soft, ash gray, with black setae interwoven, 27 mm long and 13 mm wide. The pupa is red-brown with a darker pattern. "

Mabilde's account agrees that the larva appears in October and November, but his adult emerged 10 days after pupation. He adds that the cocoon is found among dry leaves and that the larva feeds on "Cambarasinho dos campos" [*Lantana sellowiana* (Verbenaceae)] and "guaco" [*Mikania amara* (Compositae)].

The adults are very frequent in the virgin forest of Misiones almost all year-round (Jorgensen,1913) and show pompilid-like behaviour in the sun along open forest paths and ravines (Bryk, 1953).

ADULT HOST RECORDS (from Jorgensen, 1913):

Compositae:
1. *Baccharis genistelloides* Pers.
2. *Baccharis serrulata* Pers.
3. *Baccharis subopposita* DC.

4. *Baccharis tridentata* Vahl.
5. *Eupatorium kleinioides* H.B.K
6. *Eupatorium macrocephalum* Less.
7. *Eupatorium palustre* Baker
8. *Moquinia polymorpha* DC.
9. *Senecio brasiliensis* Less.
10. *Senecio icoglossus* DC.
11. *Vernonia adenophylla* Mart. ex DC.
12. *Vernonia mollissima* D. Don
14. *Vernonia polyphylla* Sch.
15. *Vernonia senecionea* Mart.

Leguminosae:
16. *Acacia riparia* H.B.K.

GEOGRAPHICAL DISTRIBUTION. Central eastern Perú, eastern Bolivia, northern Argentina, Paraguay, Uruguay, and southern Brazil.

FLIGHT PERIOD. Probably flies throughout the year, but whether it is multivoltine is not known. Collection records for adults exist for every month.

REMARKS. This species has been confused principally with *maja* F. and *lades* Cramer. *M. lades* is easily separated morphologically from *leucostigma* by the absence of vertical flanges on the uncus dorsum in males. Also, the single green spot at the base of the forewing as indicated in the original description, and a Surinam type locality could not apply to *leucostigma*.

There is no clear evidence that *maja* F. and *leucostigma* Perty are synonymous. While I agree with Forbes (1939) that *leucostigma* may be the true *maja* of Fabricius, I have retained Perty's name arbitrarily because of its more specific type locality (Brazil) and its more common usage in the literature (Zerny, 1931; Forbes, 1939) for a recognizable biological entity. The type locality for *maja* F. is "America, " which is too general to be useful, and its Fabrician description fits various species throughout the range of *Macrocneme*. Since it is the type species for the genus and the type specimen is extant, I have treated it as an indeterminant species and am relying on *leucostigma* Perty to represent the most commonly encountered species of *Macrocneme* in northern Argentina, Paraguay, and Brazil. Conceivably *leucostigma* is synonymous with one of the more narrowly endemic species from the Brazilian coast (see Table 3) but verification is impossible without the type. It is, however, easily identified, abundant, and known to have originated in Brazil. Since it is the oldest name in *Macrocneme* after *maja* and *lades* and can be tied to an identifiable species, I feel its continued use is justified.

Zerny (1931a) synonymized Draudt's form name *chionopus* under *leucostigma* Perty, following Schaus' suggestions (1924) that *lades* Cramer and *leucostigma* Perty were distinct. I am not sure this synonymy is correct. *M. leucostigma* is seldom seen with entirely white hind tarsi and it is uncommon in Perú, even though Bryk described a similar specimen from Roque, Perú in 1953 (see following). Possibly Draudt's form *chionopus* belongs in the *immanis* complex, but I see no purpose in making this transfer until the specimen referred to by Hampson is located and shown to be other than *leucostigma* Perty.

Apparently unaware that Zerny had synonymized *chionopus*, Bryk changed its rank to subspecies (1953), with the justification that entirely white hind tarsi was a subspecific character. I have examined his material (two females, Roque, Perú) and find the holotype is unquestionably *leucostigma* Perty and the allotype "male" [a female collected 11-V-1925] is clearly a different species, possibly *coerulescens* Dognin.

Draudt's form name *deceptans* refers to the same material Hampson (1898) called "Ab.1" under *M. chrysitis*. The localities were Guatemala and Rio Grande do Sul. Obviously the material was mixed, since *chrysitis* never occurs with white hind tarsi and is seldom found south of Guatemala. I was unable to locate the Guatemalan material that Hampson referred to. Most likely it has become incorporated under *lades* in the British Museum collection. The Rio Grande do Sul specimen is identifiable, and I have selected it as the lectotype for *deceptans* Draudt [=female of *leucostigma* Perty].

SPECIMENS EXAMINED (670): 343 males; 327 females. **PERU**: *Cuzco*: Cajón [hacienda], USNM. *San Martín*: Roque, NRS; Tarapoto, BMNH. "*Perú*", USNM. **BOLIVIA**: *Beni*: Río Yacuma Espíritu, 250 m, SMM. *Cochabamba*: Chapare, [upper region] Río Chipiriri, 400 m, SMM; Cochabamba, 2600 m, SMM. *Santa Cruz*: Cuatro Ojos, CM; La Junta, IML; Chiquitos, San Julian, 400 m, BMNH; Prov. del Sara, 450 m, BMNH; Prov. de Sara, Buena Vista, 750 m, BMNH, CM; Buena Vista, Ichilo, 480 m, CU; Santa Cruz, 500 m, BMNH, CAS, IML, SMM; Santa Cruz de la Sierra, BMNH, IML. *Tarija*: Río Burmejo to Río Pilcomayo, BMNH. **ARGENTINA**: *Buenos Aires*: Buenos Aires, PM; San Fernando, USNM. *Chaco*: Resistencia, IML. *Entre Ríos*: Cinco de Mayo, IML; Paraná, BMNH. *Formosa*: Formosa, BMNH; San José, N. Argentina, SMM. *Jujuy*: Calilegua, IML; Ledesma, IML; Los Perales, IML; San Pedro, 580 m, SMM; Yuto, 350 m, IML, SMM. *La Rioja*: La Rioja, BMNH. *Misiones*: Campo Viera, SMM; Caraguatay, IML; Dos de Mayo, IML, ULP; Eldorado, SMM; Iguazú, IML; Missions Territory, PM; Panambí, IML; Posadas, NRS; Pto. Aguirre, USNM. *Salta*: Aguaray, IML; Coronel Moldes, IML; Depto. San Martín, Pocitos, IML; El Morenillo, Rosario de la Frontera, IML; Orán, ULP; Salta, BMNH, NRS. *Santa Fé*: Ocampo, El Chaco, BMNH; Villa Ana, BMNH. *Tucumán*: Cadillal, IML; Depto. Río Chico, Monte Bello, IML; Depto. Trancas, Tacanas, IML; Tacanas, San P. Colalao, ULP; El Sunchal, IML; La Criolla, PM; Parque Aconquija, IML; Quebrada de Lules, IML; Siambón, IML; Tapia, 600 m, BMNH; Tucumán, 450 m, BMNH, IML, ULP, USNM; Yerba Buena, IML, PM. "Argentina" (Thomas), BMNH. **PARAGUAY**: *Asunción*: Asunción, BMNH, CM, VM. *Alto Paraná*: Puerto Bertoni, BMNH, ULP. *Caaguazú*: Caaguazú, USNM. *Central*: Nueva Italia, USP. *Cordillera*: San Bernardino, VM. *Guairá*: Independencia, SMM; Villarica, BMNH, SMM, ULP. *Itapúa*: San Luis. *Paraguarí*: Sapucay [=Sapucaí], BMNH, CM. "Central Paraguay" (Germain), BMNH; "Paraguay" (Bohls; Schrottky), BMNH. **URUGUAY**: *Paysandú*: Paysandú, CU. "Uruguay" (Pouillon), USNM. **BRAZIL**: *Bahía*: Bahía [=Salvador], (Fruhstorfer), BMNH, NRS, VM. *Distrito Federal*: Estaçao Florestal, Cabeça do Veado, 1100 m, CNC. *Mato Grosso*: Chapada dos Guimarães, Colegio Buriti, 700 m, USNM; Murtinho [railroad station], USP; Nioac [Nioaque], BMNH; Salobra, USP; Urucum, Corumbá, CU. *Minas Gerais*: Brasilia (Fruhstorfer), BMNH; Casambu [Caxambu], PM; Km 290, Rio-Bello-Barbacena, USNM; Ouro Prêto, 1100 m, CM; Passa Quatro, BMNH; Uberaba, BMNH; *Paraná*: 8 km E. Banhado, CNC; Castro, BMNH, USNM; Curitiba, PM; Iguassú [=Iguaç], BMNH; "Paraná" (Janson), BMNH; Ponta Grossa, MCZ, ULP, USP; Rolândia, CNC. *Pernambuco*: "Pernambuco" [=Recife], (Moss), BMNH. *Rio de Janeiro*: Angra dos Reis, Fazenda Japuhyba, USP; Barbacena, SMM; Ilha Grande, BMNH; Imbariê, 1200 m, SMM; Lagune de Sacuaresma [=Lagoa de Saquarema], BMNH; Maná, Itatiaia, 1200 m, CM; Nova Friburgo, PM, USNM; Organ Mts. [=Serra dos Órgãos] nr. Tijuca, BMNH; Petrópolis, USNM; Pôrto Real, BMNH, RML; "Rio de Janeiro", LACM, OX, SMM; Silvestre, CU. *Rio Grande do Sul*: Hamburgo Velho, SMM; Pelotas, CU, MCZ, SMM, USNM, USP; "Rio Grande do Sul", BMNH; Santa Maria, 1400 ft, BMNH; Santa Rosa de [Lima?], ULP. *Santa Catarina*: Hansa Humboldt, BMNH, USNM; hills between Hansa and Jaraguá, 400 m, BMNH; Jaraguá do Sul, BMNH; Joinville, IML, USNM; Mafra, 800 m, BMNH; Nova Brémen, VM; Nova Brémen, Rio Lacisz, BMNH; Nova Teutônia, USNM; "Sta. Catherines", USNM; São José, USNM. *São Paulo*: Alto da Serra, BMNH, USP; Araras, VM; Avanhandava, USP; Borhumi [=Bauru], S. Paulo, BMNH; Campo Alegre, Itirapina, 800 m, USNM; [Serra da] Cantareira, BMNH, USP; Estaçao Florestal, USP; Eugênio Léfèvre, Campos do Jordão, 1200 m, USP; Ipiranga, BMNH, USP, VM; Itaquaquecetuba, BMNH; Itanhaém, BMNH; Juquiá, Fonte Tapir, 400 m, USP; Juquiá, Km 165, 300 m, USP; Leme, VM; Marumbí, Portugal, USP; Monte Alegre, Fazenda Bom Jesús, 800 m, USP; Monte Alegre, Fazenda Sta. Maria, 1100 m, USP; Paranapanema, USNM; Piassaguera [railroad station], USP; Pôrto Cabral, Rio Paraná. USP; Rio Batalhal, BMNH; Rio Prêto, USP; São Bernardo, USP; "São Paulo", BMNH, NMG, SMM. "S. Brazil" (Boisduval), BMNH. Not located: Capta, OX.

Macrocneme megacybe Dietz,   new species

(Figs. 120-123, 177-178, 233; Map 23)

This is a member of a species complex including *M. ferrea, cyanea*, and *cupreipennis*, that is restricted to southern Brazil. The males are readily distinguished by the shape of the uncus, with its swollen dorsum, and by claspers that are asymmetrical. The females are not known with certainty. The three examples listed here are only tentatively associated on the basis of distribution. Their status may change once the sexes are better known (see Remarks).

MALE. *Head*: brownish black, occiput and vertex metallic blue; labial palpi not reaching base of antennae. *Thorax* (including legs): brownish black, disc iridescent blue with metallic scales visible through overlying hairs; patagia with mesal area largely metallic blue; tegulae with metallic streak along anteromesal margin, underside with few white scales, not visible at fringe; pectus with metallic scales on metapleuron, white spot on each pleural segment; inner mesal margin of forecoxae white, coxal grooves metallic-scaled, hind tarsi white-tipped. *Forewing*: brownish black, basal half to end of cell greenish blue, veins visible depending on angle of light, short black streak from basal angle to below cell. Underside similar to upperside, but iridescence more blue; black streak absent, retinaculum brownish black with few white scales. *Hindwing*: brownish black, discal area to base iridescent greenish blue; underside similar to upperside. *Abdomen*: iridescent green, with bright sheen more apparent on dorsum as thin stripes along lateral margins and mid-dorsally; basal tergite brownish black, with usual four white spots; pleura white-spotted on basal two segments; venter green with lateral margins iridescent; basal sternite with lateral white spots larger than those in mesal series to segment V; sternite VIII as in Fig. 132. *Genitalia*: as in Figs. 120-123 (drawn from paratype, RED prep. 39229, BMNH; 2n); dorsal processes (claspers) of valvae asymmetrical, right arm shorter and narrower at tip than left arm; dorsum of uncus inflated, with lateral margins asymmetrically expanding from base as wing-like flanges, left side larger and more open than right side; apex of juxta quadrate, not extending beyond base of ventral processes, left corner extended slightly; dorsum of aedeagus with two spines, the left prominently spinose along outer margin, the right smooth, protruding outward.

FEMALE. Length of forewing: 20 mm. Similar to the male except for minor differences in color and scale patterns. *Forewing*: basal half iridescent green, with distal margin bluish green; basal black streak extending to costa; underside iridescence bluer than above, scarcely reaching end of cell. *Hindwing*: brownish black with small iridescent green patch in distal area; underside extensively iridescent except at apex and outer margin. *Abdomen*: dorsum dark green, with lateral margins and thin mid-dorsal stripe shiny, iridescent green; pleura with two white spots basally; venter dark green with lateral margins iridescent green; small white spots in mesal series plus additional pair on basal sternite. *Genitalia*: as in Figs. 177-178 (drawn from allotype, RED prep. 267, BMNH; 1n); sternite VII of sterigma broadly u-shaped, apex broadly truncate, skewed left, caudal margin uneven, notched at center; intersegmental membrane between VI and VII with sclerotized pocket to right of mid-line; anterior margin of encircling pleurite at right reflexed to form second pocket; inner fold of lamella antevaginalis (sternite VII) as wide as overlying outer fold; lamella postvaginalis a reniform sclerite lying below the two prominent plicae and mesal protuberance of segment VIII; thickened fold forming two blind pockets in dorsal wall of ductus bursae; membrane of ductus bursae and bursa copulatrix in concentric plicae; signa the usual two opposing, scallop-shaped spiny patches.

VARIATION. Length of forewing: 20.5-21.5 mm. Wing iridescence blue, green, or mixture of the two colors, with metallic scales seldom extending beyond end of cell; forewing black streak extending to costa or remaining short and not reaching cell.

TYPE DATA. Holotype male: Alto da Serra, Santos, Brazil, 2600 ft, March 1910 (E. D. Jones), Brit. Mus. 1919-1925, BMNH. Allotype: Castro, Paraná, Brazil. Feb. 1897, (E. D.Jones), Rothschild Bequest 1939-1, BMNH. Paratypes (11): 9 males, 2 females. **BRAZIL**: *Paraná*: Castro, March (E.D. Jones), BMNH, USNM; Curitiba (Lombard), PM. *Santa Catarina*: Mafra, 800 m, December (Maller), BMNH; Sta. Catharina, [female], March, Br. Pohl, Cornell U. Lot 819, Sub 200, CU; Timbó-Blumenau, [female], ex larva, November, BMNH. *São Paulo*: Alto da Serra,

August (R. Spitz), BMNH; São Paulo, BMNH; Ypiranga, April (R. Spitz), BMNH. "Brazil" (Menestrier), PM.

BIOLOGY. No information.

GEOGRAPHICAL DISTRIBUTION. Southeastern Brazil.

FLIGHT PERIOD. Too few specimens available to accurately gauge the yearly pattern of availability. Collection records show adults in February, March, and April and again in August, November, and December.

ETYMOLOGY. The specific epithet is derived from the appearance of the dorsum of the uncus, i.e., Greek *mega* =large; Greek, f., *kybe* =head.

REMARKS. The resemblance of the female genitalia of *megacybe* to the *ferrea* holotype suggests that the two are conspecific (cf. Figs. 177 and 179). The differences in margin of the lamella postvaginalis (cf. Figs. 178 and 180) and intersegmental sclerotization between sternites VI and VII are most likely minor variations.

Because there is no direct evidence to associate males and females in *M. ferrea* and the closely related *megacybe*, it is possible this interpretation is in error. Females I have assigned to *megacybe* may represent another population of *ferrea*, a considerable range extension. Whether or not this is the case, the males of *ferrea* and the new species (*megacybe*) may have been reversed, and those presently assigned to *ferrea* actually represent the undescribed species.

Macrocneme mormo Dietz,   new species

(Figs. 128-131, 185-186, 231; Map 26)

This is another species from southern Brazil that fits the Fabrician description for *M. maja*, the type species. The medium size and restricted distribution of *M. mormo* ally it to *megacybe, bestia,* and *pelotas*. Its identification is possible only with the genitalia. In males the long, sinistrad spine on the dorsum of the aedeagus is diagnostic, differing from that in *bestia* by lacking a hooked tip. The tip can be seen exserted on dried specimens. Also, the juxta is longer and differently shaped than in *bestia*. In females the sclerite of the lamella postvaginalis, like *pelotas*, is deeply inserted and twisted beyond the ostial opening.

MALE. *Head*: brownish black, fringe of occiput with small metallic spots; labial palpi not reaching base of antennae. *Thorax* (including legs): brownish black, iridescent markings blue-green; white spots of patagia comparatively small, adjacent metallic spots about 3X as large; metallic streak on tegulae prominent, underside without white scales; pectus without metallic scales; white spots above coxae, largest on metathorax; propleuron with faint white spot; fore and mid femora tan dorsally, white below on hind femora; all tarsi blue-streaked on outer surface; hind tarsi white-tipped. *Forewing*: brownish black, basal half metallic green to blue-green, interrupted by oblique black streak from basal angle into cell, veins dark; underside similar, but black scales absent and white patch at basal angle; retinaculum white. *Hindwing*: brownish black, metallic blue in discal area; underside entirely metallic except at apex, white on costa and as basal spot on Sc+R. *Abdomen*: brownish black, interrupted by iridescent green as thin mid-dorsal and wide lateral bands, iridescence becoming blue-green caudally; venter with iridescent scales laterally; mesal series of white spots diminishing in size caudally, basal sternite mostly white. *Genitalia*: as in Figs. 128-131 (drawn from paratype, RED prep. 39233, USNM; 3n); dorsal processes (claspers) of valvae slightly asymmetrical, left arm longer, less curved to apex than right arm; uncus not skewed, dorsum prominently round with lateral margins as horizontal, subequal flanges; juxta long, reaching well beyond base of ventral processes, apex broadly incised at right, tip margined with short spines; dorsum of aedeagus with two spines, the left approximately 4X length of right, curving sinistrad, the right straight, slender.

FEMALE. Length of forewing: 17.5 mm. Essentially like male, except iridescent markings in allotype appear more blue than in holotype. *Thorax*: pectus less white; legs brownish black, with tan and white absent from femora and from above coxae. *Genitalia*: as in Figs. 185-186 (drawn from allotype, RED prep. 39236, USNM; 1n); sternite VII of sterigma U-shaped, apex cephalad with margin skewed left and narrowly continuous with sclerotized pocket invaginated between sternites VI and VII; segment VII (tergite and sternite) and posterior half of VI more strongly sclerotized than anterior segments; sclerite of lamella postvaginalis large, bent, well inserted beyond ostial opening, left margin twisted inward within ductus bursae; dorsal wall of ductus bursae with two thickened plicae; signa two opposing C-shaped patches with erect spines.

VARIATION. Length of forewing: 16.0-18.0 mm. Iridescence on wings or body green, blue-green, or blue; underside of tegulae occasionally white (Bot. Garden, Rio), but not visible on mesal fringe; black streak on forewing occasionally reduced to few scales (Rio), making iridescence appear entire rather than interrupted; abdominal iridescence usually evident in thin longitudinal bands, but occasionally suffusing over entire dorsum.

TYPE DATA. Male holotype: Rio Mucuri, BR 101, Mun. Mucuri, Bahia, Brazil, 5-V-69 (S.S. Nicolay), USNM Type# 73265, USNM. Allotype: data similar to male (no type no.). Paratypes (45): 34 males, 11 females. **BRAZIL**: *Bahía*: BR 101, Rio Mucuri, Mun. Mucuri, May (Nicolay), USNM. *Rio de Janeiro*: Angra dos Reis, Fazenda Japuhyba, July, September, October (Travassos F°), USP; Botanical Garden, Rio, May, USNM; Imbariê, 1200 m, April, July (Ebert), SMM; Petrópolis (Foetterle), VM; Rio de Janeiro, October, November (Arp; Schott; Smith), BMNH, CM, USNM, VM; Santa Ana, nr. Rio de Janeiro, NMG; Sarapyhy [=Sarapuí], Rio (Foetterle), VM. *Rio Grande do Sul*: Pelotas (Biezanko), CU; Porto Alegre (Foetterle), VM. *Santa Catarina*: "South Brazil, St. Catharina", BMNH. *São Paulo*: [Serra da] Cantareira, April (Spitz), BMNH; Juquiá, Fazenda Poco Grande, October, USP; Osasco, USP; Porto Cabral, Rio Paraná, October (Travassos F°), USP; Ubatuba, sea level, CM; Ypiranga, March (Travassos F°), USP.

The localities Cantareira, Porto Cabral, and Ypiranga were taken from unassociated females and should be considered tentative until males of *mormo* are taken from the same localities.

BIOLOGY. No information.

GEOGRAPHICAL DISTRIBUTION. Known only from southeastern Brazil.

FLIGHT PERIOD. Adult collection records are available for all months except February, June, August, and December. Probably flies throughout the year.

ETYMOLOGY. The specific epithet is taken from the Greek *mormo,* meaning hobgoblin, used to frighten children into good behavior.

REMARKS. The presence of this species can be disguised when the phenotype appears to mimic that of a sympatric species. In what appeared to be a homogeneous series (ex CU) of *M. pelotas* from Pelotas, Brazil, there was one male of *mormo.* The same phenomenon is found among selected populations in the *semiviridis* complex in Colombia.

Macrocneme pelotas Dietz, new species

(Figs. 112-115, 183-184, 225; Map 16)

Like other *Macrocneme* from southern Brazil, this species is easily confused with its congeners unless the genitalia are examined. In males the round, non-flanged uncus, the three-pronged aedeagus, and the peculiarly shaped juxta suggest a relationship with *M. thyridia,* even though the two species are widely allopatric. In females the configuration of the sterigma including the large inter-segmental pocket and the deeply inserted sclerite of the lamella postvaginalis, suggest a relationship closer to *mormo* than to other sympatric species.

MALE. *Head*: brownish black, occiput with two small metallic blue spots; labial palpi not reaching base of antennae. *Thorax* (including legs): brownish black; iridescent markings bright blue; patagia with foremargin entirely metallic blue rather than as separate spots; tegulae metallic-streaked from shoulder to middle of sclerite, underside lacking white scales; pectus hairy; white spots only on propleuron, above midcoxae, and on metepisternum; coxal grooves with scattered metallic scales; forecoxae and all tibiae thinly streaked with blue, hind tarsi white-tipped. *Forewing*: dark brownish black, basal half bright metallic blue interrupted basally by oblique black streak from basal angle to costa; diffuse white spot at humeral angle; underside similarly metallic at base, but black scales absent; retinaculum white and brownish black. *Hindwing*: dark brownish black, discal area with metallic blue patch mostly below cell; underside metallic blue to beyond cell, apex brownish black. *Abdomen*: dorsum brownish black with iridescent blue as thin mid-dorsal stripe and as wider lateral bands; pleura with single white spot basally; venter brownish black with lateral margins thinly blue; two white spots laterally on sternite I, smaller mesal spots on sternites II-V. *Genitalia*: as in Figs. 112-115 (drawn from paratype, RED prep. 39228, USNM; 2n); dorsal processes (claspers) of valvae asymmetrical, right arm more curved and slightly shorter than left arm, tip of arms round with inner surface convex at tip; mesal sclerite prominently knobbed; dorsum of uncus distinctly round without lateral flanged margins, two short vertical flanges basad; juxta deeply incised with left side elongate, tip round, spined, right side short, tip sharp; inner membrane with ovoid patch of fine spines at lower left; aedeagus with two unusual spines pointing outward from dorsum, left spine with secondary, thorn-like spine on inner margin, right spine with tip hooked.

FEMALE. Length of forewing: 20 mm. Very similar to male except for minor variation in iridescent scaling. *Thorax*: metallic spots on patagia reduced to tiny lateral dots; metallic streaks absent from forecoxae and hind tibiae. *Abdomen*: pleura with three white spots, one basad and two distad (segments VI and VII); venter with six well-defined white spots in mesal series. *Genitalia*: as in Figs. 183-184 (drawn from allotype, RED prep. 274, USNM; 3n); sternite VII of sterigma shield-shaped, apex skewing slightly to left, narrow sclerotized strip with invaginated intersegmental pocket at right, lateral flaps asymmetrical, caudal margin uneven; inner fold of lamella antevaginalis incised at left; sclerite of lamella postvaginalis ovoid, bent at ostium and deeply inserted, left margin twisted inward within ductus bursae; dorsal wall of ductus bursae without accessory fold; signa of two opposing C-shaped spiny patches.

VARIATION. Length of forewing: 17.2-19.3 mm. Color of iridescence varies, bright blue in populations from Pelotas and green in those from Itatiaya or Petrópolis.

TYPE DATA. Male holotype: Pelotas, Rio Grande do Sul, Brazil, 25-I-52, C.M. de Biezanko, no. CB-2060, USNM Type# 73266, USNM. Allotype: data similar to male (no type no.), USNM. Paratypes (52): 38 males, 14 females. **BRAZIL**: *Minas Gerais*: São Lourenço (Foetterle), VM; Serra de Casperaó (Knudsen), NRS. *Paraná*: Ponta Grossa, April (Richards), BMNH. *Rio de Janeiro*: Angra dos Reis, Fazenda Japuhyba, July-October (Travassos F°), USP; Itatiaya, 800 m, April (Ebert), SMM; Petrópolis (Schaus), USNM. *Rio Grande do Sul*: Pelotas, December-April, July (Biezanko), CM, CU, IML, MCZ, SMM; "Rio Grande do Sul", PM. *Santa Catarina*: Jaraguá (Hoffmann), VM. *São Paulo*: Alto da Serra [Santos], April, July, November (Spitz), BMNH, USP; Itanhaém, December (Munroe; Spitz), CNC, USP; Juquiá, Fazenda Poco Grande, October (C.D.Z.), USP.

BIOLOGY. No information.

GEOGRAPHICAL DISTRIBUTION. Restricted to southeastern Brazil.

FLIGHT PERIOD. Probably active throughout the year. Adult collection records are available for all months except May and June.

ETYMOLOGY. This species is named for the locality where it was commonly collected by Professor Biezanko.

REMARKS. This species has often been mistakenly identified as *M. coerulescens* Dognin, but its distribution makes this identification impossible.

M. pelotas is one of three species (also *ferrea* and *cyanea*) that has a patch of spines on the inner membrane of the juxta. As these are all southern Brazilian species, this character may indicate a degree of phyletic affinity that might otherwise be unsuspected.

Any series that is presumed to be *pelotas* should be examined carefully for examples of sympatric species. Among a series of homogenous specimens from Pelotas I found one example of *mormo* that was identical to the bright blue examples of *pelotas*.

SPECIES TRANSFERRED
FROM *MACROCNEME*

The following three species were presumably placed in *Macrocneme* when the type of *Poliopastea, P. plumbea* Hampson was transferred to *Macrocneme* by Fleming (1957). Their taxonomic placement is open to question, but none should be considered in *Macrocneme*. From a brief examination of their genitalia, I would suggest the following changes:

1. *P. vidua* Bryk, 1953 = *Dycladia marmana* Schaus, 1924 [NEW SYNONYMY]. This species also appears allied to species in *Ichoria* near *tricincta* Herrich-Shaffer or *chalcomedusa* Druce.
2. *P. viridis* (Druce), 1883 = *Pseudosphenoptera viridis* (Druce), new combination.
3. *P. pava* (Dognin), 1893 = *Chrysocale pava* (Dognin), new combination.

CHECKLIST OF SPECIES
with known distributions

Genus *Macrocneme* Hübner, 1818

1. *adonis* Druce, 1884   Mexico to Brazil
    chiriquicola Strand, 1917
    cinyras Schaus, 1889
2. *ancaverdia* Dietz, n. sp.   Venezuela to Bolivia
3. *aurifera* Hampson, 191   Trinidad to Peru & Brazil
    spinivalva Fleming, 1957
4. *bestia* Dietz, n.sp.   Brazil
5. *bodoquero* Dietz, n. sp.   Colombia, Peru, Brazil
6. *cabimensis* Dyar, 1914   Central America, Colombia, Ecuador
7. *chrysitis* (Guérin, 1844)   Mexico, Central America
8. *coerulescens* Dognin, 1906   Venezuela to Peru
    yepezi Förster, 1950
9. *cupreipennis* Walker, 1856   Brazil, Argentina
10. *cyanea* (Butler, 1876)   Brazil, Argentina
11. *durcata* Dietz, n. sp.   Venezela to Bolivia & Brazil
12. *ferrea* Butler, 1876   Brazil
13. *habroceladon* Dietz, n. sp.   Colombia to Bolivia
14. *imbellis* Dietz, n. sp.   Peru
15. *immanis* Hampson, 1898   Peru to Argentina
16. *iole* Druce, 1884   Nicaragua to Colombia
17. *lades* (Cramer, [1776])   Mexico to Brazil
    aurata (Walker, 1854)
18. *leucostigma* (Perty, 1833)   Brazil, Argentina, Peru
    chionopus Draudt, 1916
    deceptans Draudt, 1916
19. *maja* (F., 1787)   ? Brazil
20. *megacybe* Dietz, n. sp.   Brazil
21. *melanopeza* Dietz, n. sp.   Peru
22. *mormo* Dietz, n. sp.   Brazil
23. *oponiensis* Dietz, n. sp.   Colombia, Ecuador

24. *orichalcea* Dietz, n. sp.   Venezuela to Bolivia
25. *pelotas* Dietz, n. sp.   Brazil
26. *semiviridis* Druce, 1911   Venezuela, Colombia, Ecuador
27. *tarsispecca* Dietz, n. sp.   Colombia to Bolivia
28. *thyra* Möschler, 1883   Panama to Brazil
    affinis Klages, 1906
    albiventer Dognin, 1923
    boliviana Draudt, 1916
    chlorota (Dognin, 1914)
    intacta Draudt, 1916
29. *thyridia* Hampson, 1898   Guianas to Bolivia
    euphrasia Schaus, 1924
    guyanensis Dognin, 1911
30. *zongonata* Dietz, n. sp.   Brazil, Peru, Bolivia

Appendix

Artificial Diet for Rearing
Ctenuchid and Arctiid Larvae (Lepidoptera)*

Ingredients:

250	g.	dried beans (lima, pinto, etc.)
50	g.	Brewer's yeast
25	g.	raw wheat germ
5	g.	ascorbic acid
5	g.	methyl paraben
1.5	g.	potassium sorbate
.055	g.	tetracycline hydrochloride
15	g.	agar
2.5	ml.	formaldehyde (35%)
1000	ml.	distilled water
1.65	g.	linseed oil (.01g/100 wt. or 58 drops)

Procedure:

1. Weigh all dry ingredients.
2. Soak beans overnight. Drain.
3. Heat 600 ml. water to boiling. Add beans and allow water to just boil. Remove from heat.
4. Drain water into measured container. Add additional water to 600 ml. Cool to 55°C.

5. In a blender, mix all the dry ingredients except agar with the 600 ml. water. Blend 10-20 seconds and add linseed oil and formaldehyde.
6. Add beans gradually and blend until smooth, approximately 5-6 minutes.
7. Dissolve agar in remaining 400 ml. water. Cool to 55°C and add to bean mixture. Blend until well mixed.
8. Spoon or pour immediately into containers. Allow to cool to room temperature (about one hour). Cover and refrigerate in plastic bags.

* Modified from Shorey and Hale (1965) and Schroeder (1970).

Literature Cited

Beebe, W. 1955. Two little-known selective insect attractants. *Zoologica* (N.Y.) 40(2):27-32.

— and H. Fleming 1951. Migration of day-flying moths through Portachuelo Pass, Rancho Grande, north-central Venezuela. *Zoologica* (N.Y.) 36(19):243-254, 1 pl.

— and R. Kenedy 1957. Habits, palatability, and mimicry in thirteen ctenuchid moth species from Trinidad, B.W.I. *Zoologica* (N.Y.) 42(4):147-158, 2 pls.

Bernardi, N. 1973. Warningly-colored pupae of neotropical ctenuchid moths (Lep., Ctenuchidae). *Rev. Bras. Entomol.* 17(14):105-108.

Blest, A. D. 1964. Protective display and sound production in some New World arctiid and ctenuchid moths. *Zoologica* (N.Y.) 49:161-181.

— T.S. Collett, and J. D. Pye 1963. The generation of ultrasonic signals by a New World arctiid moth. *Proc. R. Soc. Lond., B. Biol. Sci.* 158:196-207.

Boppré, M. 1978. Chemical communication, plant relationships, and mimicry in the evolution of danaid butterflies. *Entomol. Exp. Appl.* 24:264-277.

Brown, F. M. 1941. A gazetteer of entomological stations in Ecuador. *Ann. Entomol. Soc. Am.* 34:809-851.

Brown, K. S., Jr. 1977. Geographical patterns of evolution in Neotropical forest Lepidoptera (Nymphalidae: Ithomiinae and Nymphalinae-Heliconiini), *in* H. Descimon, (ed.), Biogéographie et evolution en Amérique Tropicale, pp. 118-160.

— 1979. *Ecologia geográfica e evolução nas florestas neotropicais.* 265 pp.; *Apêndices,* 120 pp. Campinas: Univ. Estadual de Campinas [published thesis].

— and O. H. H. Mielke 1972. The Heliconians of Brazil (Lepidoptera: Nymphalidae). Part 2. Introduction and general comments, with a supplementary revision of the tribe. *Zoologica* (N.Y.) 57:1-40.

—, P.M. Sheppard, and J.R.G. Turner 1974. Quaternary refugia in Tropical America: Evidence from race formation in *Heliconius* butterflies. *Proc. R. Soc. Lond., B. Biol. Sci.* 187:369-378.

Bryk, F. 1953. Lepidoptera aus dem Amazonasgebiete und aus Peru gessammelt von Dr. Douglas Melin und Dr. Abraham Roman. *Ark. Zool.* 5(1):1-268.

Burmeister, H. 1878. *Description Physique de la République Argentine*, vol. 5: Lépidoptères. 524 pp. Buenos Aires.

Butler, A.G. 1876. Notes on the Lepidoptera of the family Zygaenidae, with descriptions of new genera and species. *J. Linn. Soc. Lond., Zool.* 12:342-407.

Cevallos, M. A. 1957. El "caterpilar" (*Ceramidia viridis*) del banano. *Cien. y Nat.* (Rev. Instit. Cien. Nat. Univ. Cent. Quito) 1(1):13-14.

Costa Lima, A. da 1950. *Insectos do Brazil*, Vol. 6: Lepidópteros. 420 pp. Rio de Janeiro: Esc. Nac. Agron.

— 1968. *Quarto Catalogo dos Insetos que Vivem nas Plantas do Brazil, seus Parasitos e Predatores*, vols. 1-4. Rio de Janeiro: Ministerio Agricul.

Cramer, P. [1775-1776]. *De Uitlandsche Kapellen voorkomende in de drie waereld-deelen Asia, Africa, en America*, vol. 1. 156 pp., 96 pls. Amsterdam and Utrecht.

Descimon, H. (editor) 1977. Biogéographie et evolution en Amérique Tropicale. *Publ. Lab. Zool., Ecole Norm. Sup.*, 9. 344 pp.

Dietz, R. E., IV, and W. D. Duckworth 1976. A review of the genus *Horama* Hübner and the reestablishment of the genus *Poliopastea* Hampson (Lepidoptera: Ctenuchidae). *Smithson. Contrib. Zool.* no. 215. 53 pp.

Dognin, P. 1893. Lepidoptères nouveaux de l'Amérique du Sud, principalment de Loja et environs (Equateur). *Ann. Soc. Entomol. Belg.* 37(7):367-374.

— 1906. Hétérocères nouveaux de l'Amérique du Sud. *Ann. Soc. Entomol. Belg.* 50:178-186.

— 1911. *Hétérocères nouveaux de l'Amerique du Sud*, fasc. 2. 56 pp. Rennes.

— 1914. *Hétérocères nouveaux de l'Amerique du Sud*, fasc. 7. 32 pp. Rennes.

— 1923. *Hétérocères nouveaux de l'Amerique du Sud*, fasc. 23. 34 pp. Rennes.

Doyen, J. T., and R. E. Somerby 1974. Phenetic similarity and Müllerian mimicry among darkling ground beetles (Coleoptera: Tenebrionidae). *Can. Entomol.* 106:759-772.

Draudt, M. [1916-1919]. Family Syntomidae, *in* A. Seitz (ed.), *The Macrolepidoptera of the World*, vol. 6, pp. 37-230. Stuttgart: A. Kernen [English edition; for dates, see Griffen, 1936].

Druce, H. 1881-1900. *Biologia Centrali-Americana: Insecta; Lepidoptera; Heterocera.* 3 vols.

— 1883. Descriptions of new species of Zygaenidae and Arctiidae. *Proc. Zool. Soc. Lond.*, pp. 372-384.

— 1911. Descriptions of some new species of Heterocera from tropical South America, and two new species of Geometridae from West Africa. *Ann. Mag. Nat. Hist.*, ser. 8, vol. 7(39):287-294.

Dunning, D.C. 1968. Warning sounds of moths. *Z. Tierpsychol.* 25: 129-138.

Dyar, H. G. 1914. Report on the Lepidoptera of the Smithsonian Biological Survey of the Panama Canal Zone. *Proc. U.S. Nat. Mus.* 47(2050):139-350.

Emsley, M. 1964. The geographical distribution of the color-pattern components of *Heliconius erato* and *Heliconius melpomene* with genetical evidence for the systematic relationship between the two species. *Zoologica* (N.Y.) 49(3):245-286, 2 pls.

Eyre, S. R. 1968. Vegetation and Soils, a World Picture, 2nd edition. xvi+ 328 pp. London: E. Arnold.

Fabricius, J. C. 1781. *Species insectorum....* vol. 2. 517 pp. Hamburg and Cologne: E. Bohnii.
— 1787. *Mantissa insectorum....*vol. 2. 382 pp. Hafniae [=Copenhagen]: C.G. Proft.

Fairchild, G.B., and C.O. Handley, Jr. 1966. Gazetteer of collecting localities in Panama *in* R.L. Wenzel and V.J. Tipton (eds.), *Ectoparasites of Panama*, pp. 9-20, 1 map. Chicago: Field Museum of Natural History.

Felder, C. and R. Felder 1862. Specimen faunae lepidopterologicae riparum fluminis Negro superioris in Brasilia septentrionali. *Wien. Entomol. Monatsschr.* 6:65-80, 109-126, 229-235.

Field, W.D. 1975. Ctenuchid moths of *Ceramidia* Butler, *Ceramidiodes* Hampson, and the Caca species group of *Antichloris* Hübner. *Smithson. Contrib. Zool.* no. 198, 45 pp.

Fleming, H. 1957. The Ctenuchidae (Moths) of Trinidad, B.W.I. Part I. Euchromiinae. *Zoologica* 42(3):105-130, 3 pls.

Forbes, W.T.M. 1939. The Lepidoptera of Barro Colorado Island, Panama. *Bull. Mus. Comp. Zool., Harv. Univ.* 85(4):vii+ 97-322, 8 pls.

Förster, W. 1950. Liste der von Pater Cornelius Vogl in Maracay und Caracas gesammelten Schmetterlinge. III. Syntomidae. *Bol. Entomol. Venez.* 8(1-2):43-67, 1 pl.

Griffen, F.J. 1936. The contents of the parts and the dates of appearance of Seitz' *The Macrolepidoptera of the World*, vols. 1 to 16, 1907-1935. *Trans. R. Entomol. Soc. Lond.* 85(10):243-279.

Guérin-Méneville, F.E. 1844. *Iconographie du règne animal de G. Cuvier... Insectes.* 576 pp., 110 pls. Paris: J. B. Baillière [*Les Lépidoptères*, pp. 466-530, pls.76-91].

Haffer, J. 1979. Quaternary biogeography of tropical lowland South America *in* W.E. Duellman, (ed.), The South American Herpetofauna: Its origin, evolution, and dispersal, pp. 107-140. *Monogr. Mus. Nat. Hist., Univ. Kansas*, no. 7.

Hagmann, G. 1938. Syntomideos (Amatideos=Euchromideos) do Estado do Pará *in* Livro Jubilar do Professor Lauro Travassos, editado para commemorar o 25° anniversario de suas actividades scientificias (1913-1938), pp. 185-194. Rio de Janeiro: Instit. Oswaldo Cruz.

Hampson, G.F. 1898. *Catalogue of the Lepidoptera Phalaenae in the British Museum*, vol. 1:xxii+559 pp, 17 pls. London.

— 1914. *ibid.*, Supplement, vol. 1:xxviii+858 pp, 41 pls. London.

Harrison, J.O. 1959. Notes on the life history and habits of *Ceramidia butleri* Möschler, a pest of bananas (Lepidoptera: Syntomidae). *Ann. Entomol. Soc. Am.* 52:351-354.

Herrich-Schäffer, G.A.W. 1850-[1869]. *Sammlung neuer oder wenig bekannter aussereuropäischer Schmetterlinge*, vol. 1: 84 pp, 120 pls; vol. 2:4 pp, 8 pls. Regensburg: G.J. Manz.

Hübner, J. [1808-]1818. *Zuträge zur Sammlung exotischer Schmettlinge* [*sic*]. Erstes Hundert. Augsburg: J. Hübner [pl. 12 dated 1809-1813].

— 1816-[1826]. *Verzeichniss bekannter Schmettlinge [sic]*. 431 pp. Appendix, 72 pp. Augsburg.

Jorgensen, P. 1913. Zur Kenntnis der Syntomiden Argentiniens (Lep.). *Z. Wiss. Insektenbiol.* 9:3-7, 33-37, 74-77.

Kaye, W. J. 1913. A few observations in mimicry. *Trans. [R.] Entomol. Soc. Lond.* 61(1):1-10, 1 pl.

Kerr, J.G. 1910. Remarks upon the Zoological Collection of the University of Glasgow, made on the occasion of the visit by the Natural History Society, on March 12, 1910. *Glas. Nat.* 2(4):97-111.

Kettlewell, H.B.D. 1959. Brazilian insect adaptations. *Endeavour* 18(72):200-210, 24 figs.

Kirby, W.F. 1892. *A Synonymic Catalogue of Lepidoptera Heterocera (Moths)*. vol. 1: Sphinges and Bombyces. xii+951 pp. London.

Kiriakoff, S.G. 1948. Recherches sur les organes tympaniques des Lépidoptères en rapport avec la classification. *Bull. et Ann. Soc. Entomol. Belg.* 84:231-276.

Klages, E.A. 1906. On the syntomid moths of southern Venezuela collected in 1898-1900. *Proc. U.S. Nat. Mus.* 29(1434):531-552.

Lamas, M.G. 1976. A gazetteer of Peruvian entomological stations (based on Lepidoptera). *Rev. Peru. Entomol.* 19(1):17-25.

Lemaire, C. 1977. Biogéographie des Attacidae de l'Equateur (Lepidoptea) *in* H. Descimon, (ed.), *op. cit.*, pp. 223-306.

Mabilde, A.P. 1896. *Guia practica para os principiantes colleccionadores de insectos contendo a descripção fiel de perto de 1000 borboletas.... Estudo sobre a vida de insectos do Rio Grande do Sul e sobre a caça, classificação e a conservação....* 238 pp. Pôrto Alegre: Gundlach and Schuldt.

Möschler, H.B. 1878. Beiträge zur Schmetterlings-Fauna von Surinam, vol. 2. *Verh. Zool.-Bot. Ges. Wien* 27:629-700, Taf. VIII-X.

— 1883. Beiträge zur Schmetterlings-Fauna von Surinam, vol. 5 (supplement) *Verh. Zool.-Bot. Ges. Wien* 32:303-362, Taf. XVII-XVIII.

Moss, A.M. 1947. Notes on the Syntomidae of Pará with special reference to wasp mimicry and fedegoso, *Heliotropium indicum* (Boraginaceae), as an attractant. *Entomologist* (Lond.) 80:30-35.

Perty, J.A.M. 1833. *Delectus animalium articulatorum quae in itinere per Brasiliam annis 1817-1820....* 224 pp, 40 pls. Monachii [=Munich] [*Lepidoptera*, pp. 151-164, pls. 29-32].

Pliske, T.E. 1975. Attraction of Lepidoptera to plants containing pyrrolizidine alkaloids. *Environ. Entomol.* 4(3):455-473.

Rothschild, M. 1973. Secondary plant substances and warning coloration in insects. *Symp. R. Entomol. Soc. Lond.* 6:59-83.

— , J. von Euw, and T. Richstein 1973. Cardiac glycosides (heart poisons) in the polka-dot moth, *Syntomeida epilais* Walker (Ctenuchidae:Lep.), with some observations on the toxic qualities of *Amata(=Syntomis) phegea* (L.). *Proc. R. Soc. Lond., B. Biol. Sci.* 183:227-247.

— , R.T. Aplin, P.A. Cockrum, J.A. Edgar, P. Fairweather, and R. Lees 1979. Pyrrolizidine alkaloids in arctiid moths (Lep.) with a discussion on host plant relationships and the role of these secondary plant substances in the Arctiidae. *Biol. J. Linn. Soc.* 12:305-326, 2 figs.

Schaus, W. 1889. Descriptions of new species of Mexican Heterocera. *Entomol. Am.* 5:87-90.

— 1924. New species of moths in the United States National Museum. *Proc. U.S. Nat. Mus.* 65(2520):1-74.

Schroeder, W.J. 1970. Rearing the pecan bud moth on artificial diet. *J. Econ. Entomol.* 63:650-651.

Schrottky, C. 1909. "Mimetische" Lepidopteren ein Beitrag zur Kenntnis der Syntomidae Paraguays. *Dtsch. Entomol. Z., "Iris"* (Dresden) 22(2-3):122-132.

Seitz, A. 1890. Die Schmetterlingswelt des Monte Corcovado. *Entomol. Zeitung* (Stettin) 51(7-12):258-266.

— 1916. Syntomidae: General topics *in* A. Seitz, (ed.), *The Macrolepidoptera of the World*, vol. 6, pp. 33-37 [English edition; for date, see Griffen, 1936].

Selander, R.B., and P. Vaurie 1962. A gazetteer to accompany the "Insecta" volumes of the "Biologia Centrali-Americana." *Am. Mus. Novit.* 2099:1-70, 8 figs.

Shorey, H.H., and R.L. Hale 1965. Mass-rearing of the larvae of nine noctuid species on a simple artificial medium. *J. Econ. Entomol.* 58:522-524.

Simpson, B.B. 1979. Quaternary biogeography of the high montane regions of South America, *in* W. E. Duellman, (ed.), The South American Herpetofauna: its origin, evolution, and dispersal, pp. 157-188. *Monogr. Mus. Nat. Hist. Univ. Kansas*, no. 7.

—, and J. Haffer 1978. Speciation patterns in the Amazonian forest biota. *Ann. Rev. Ecol. Syst.* 9:497-518.

Strand, E. 1917. Neue und wenig bekannte Nebenformen von Syntomididen. *Arch. Naturgesch.* 82A(2):79:86.

Turner, J.R.G. 1971. Studies of Müllerian mimicry and its evolution in burnet moths and heliconid butterflies, *in* R. Creed (ed.), *Ecological Genetics and Evolution: Essays in Honour of E. B. Ford*. New York: Appleton-Century-Crofts.

— 1977. Forest refuges as ecological islands: Disorderly extinction and the adaptive radiation of Müllerian mimics, *in* H. Descimon, (ed.), *op. cit.*, pp. 98-117.

Vanderzant, E.S., D. Kerur, and R. Reiser 1957. The role of dietary fatty acids in the development of the pink bollworm. *J. Econ. Entomol.* 50(5):606-608.

Walker, F. 1854. *List of the specimens of lepidopterous insects in the collection of the British Museum. Lepidoptera-Heterocera*, 1:1-278. London.

— 1856. *Ibid.*, 7:1509-1808.

Watson, A. 1975. A reclassification of the Arctiidae and Ctenuchidae formerly placed in the thyretid genus *Automolis* Hübner (Lepidoptera), with notes on warning coloration and sound. *Bull. Br. Mus. (Nat. Hist.) Entomol. Suppl.* 25:[1]-104.

Zerny, H. 1912. Syntomidae, *in* H. Wagner, (ed.), *Lepidopterorum Catalogus*, Part 7, 179 pp. Berlin: W. Junk.

— 1931a. Beiträge zur Kenntnis der Syntomiden. *Dtsch. Entomol. Z., "Iris"* (Dresden) 45:1-27, 1 pl.

— 1931b. Ergebnisse einer zoologischen Sammelreise nach Brasilien, insbesondere in das Amazonasgebiet, ausgeführt von Dr. H, Zerny. VII Teil, Lepidoptera III: Die Syntomiden des Staates Para. *Ann. Naturhist. Mus. Wien* 45:225-263.

Maps and Figures

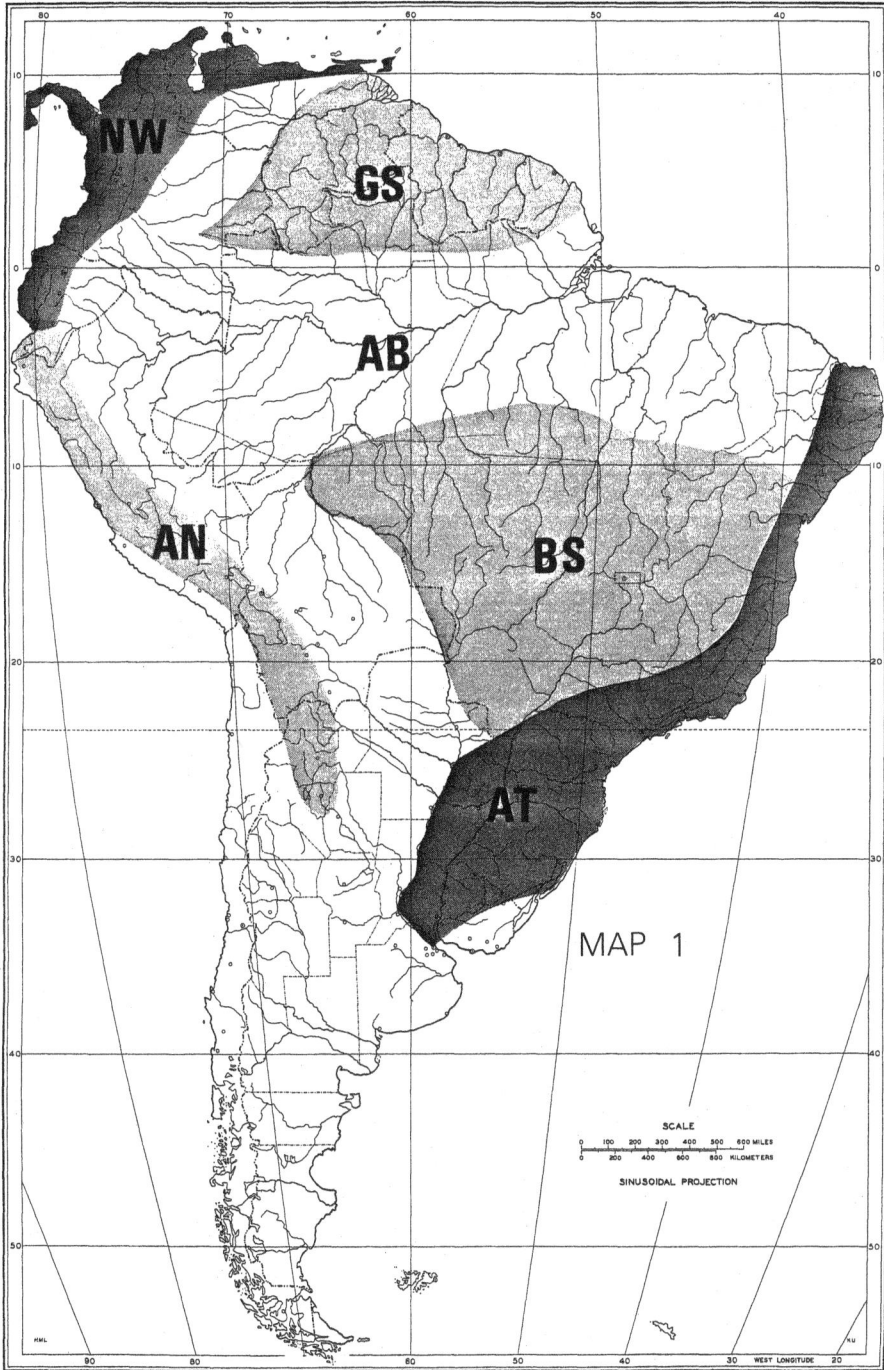

Map 1. Biogeographial Regions

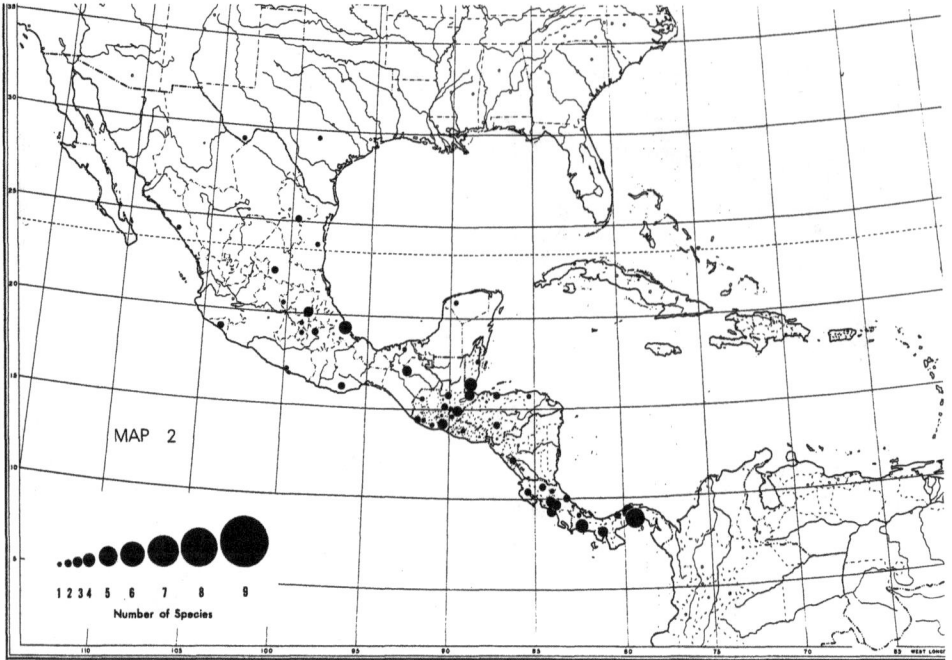

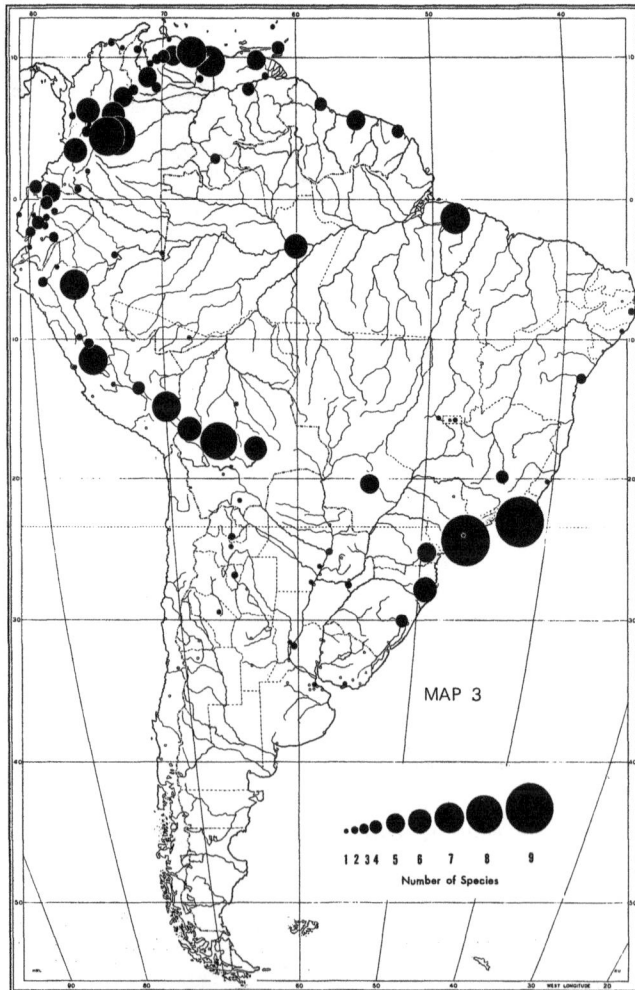

Maps 2 and 3. Sympatry in *Macrocneme*

MAP 4

adonis (1)

SCALE

CONIC PROJECTION

MAP 5

● adonis (2)

SCALE

SINUSOIDAL PROJECTION

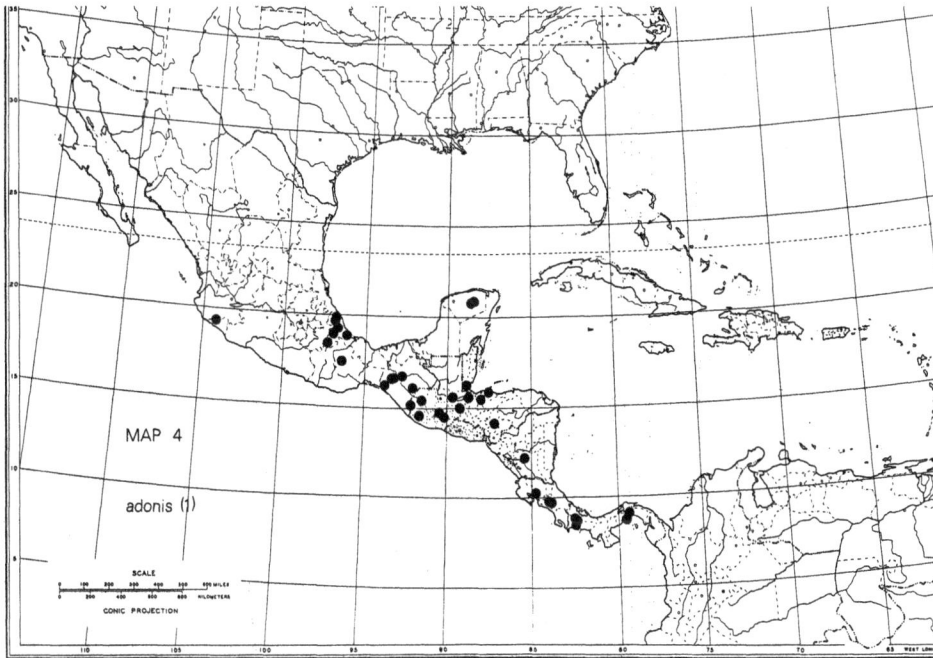

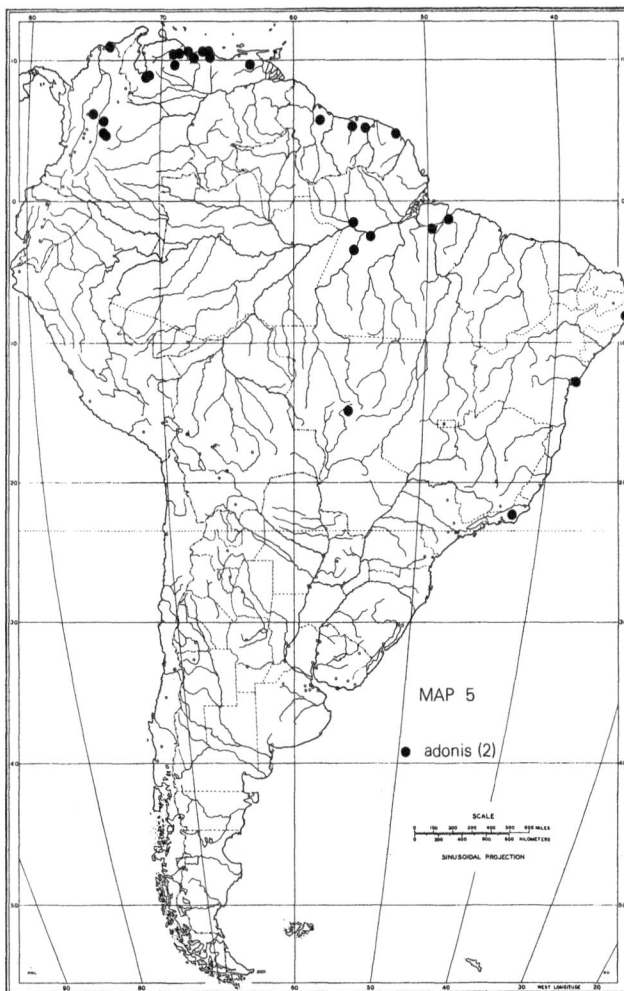

Maps 4 and 5. *M. adonis*

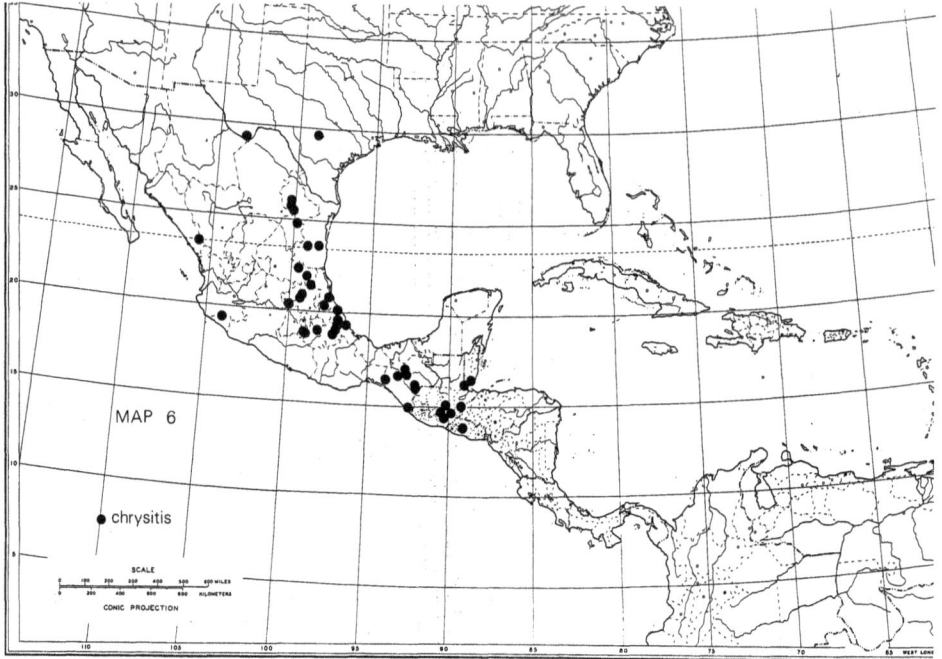

Map 6. *M. chrysitis*

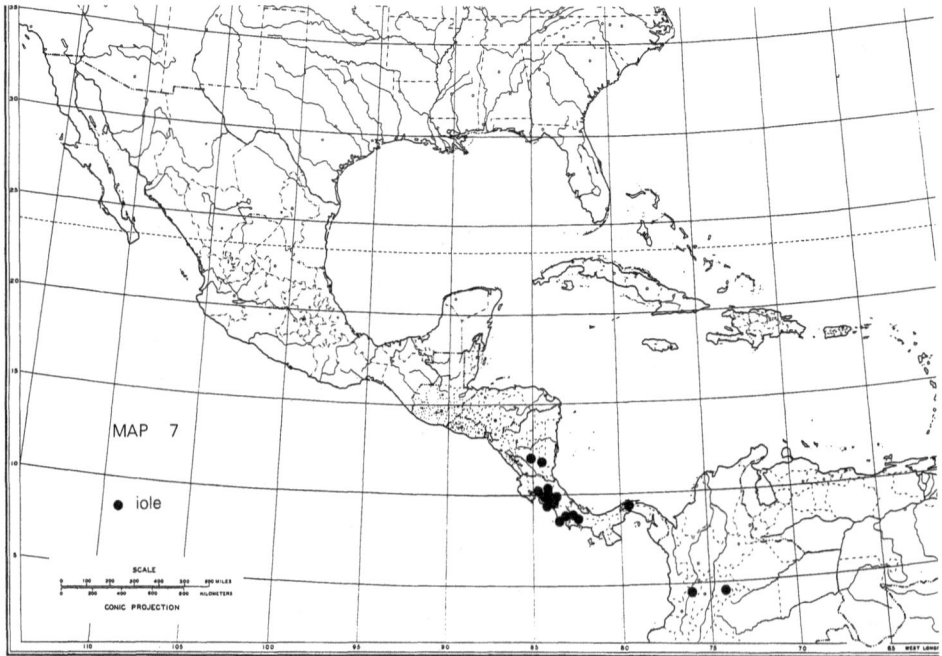

Map 7. *M. iole*

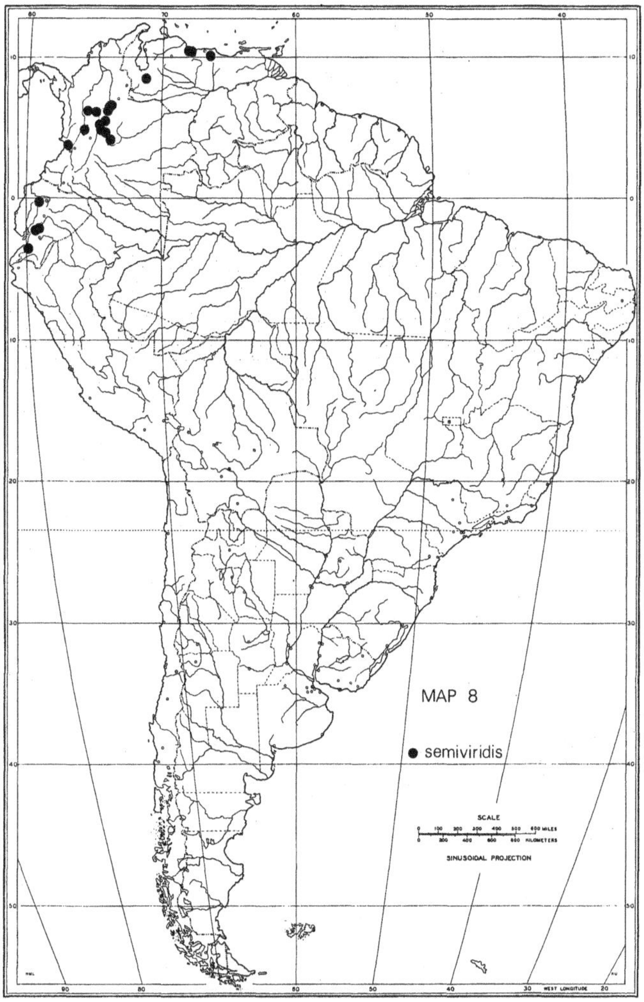

MAP 8

● semiviridis

SCALE

SINUSOIDAL PROJECTION

Map 8. *M. semiviridis*

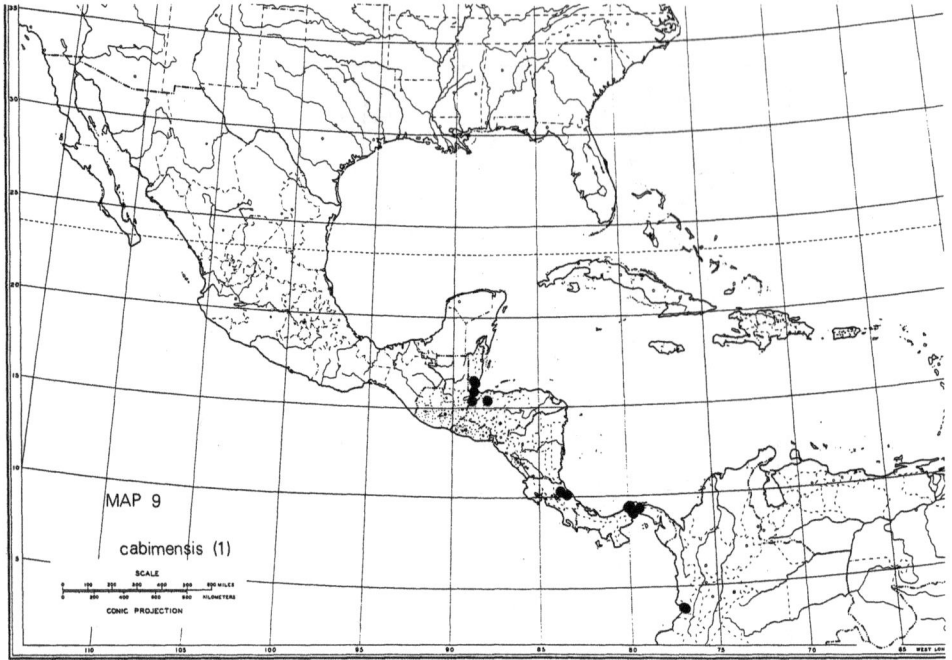

Maps 9 and 10. *M. cabimensis*

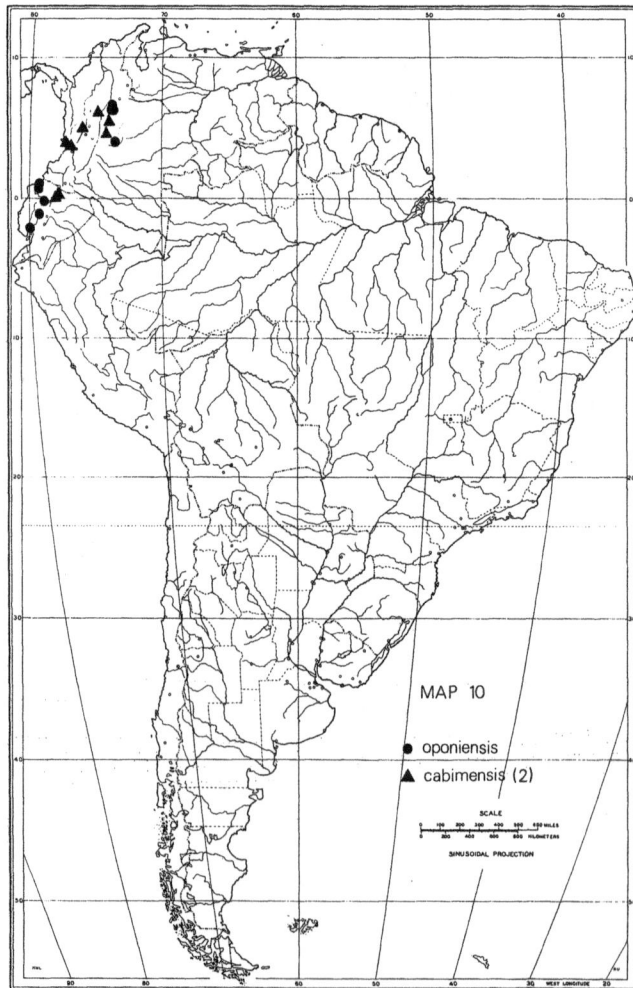

Map 10. *M. oponiensis*

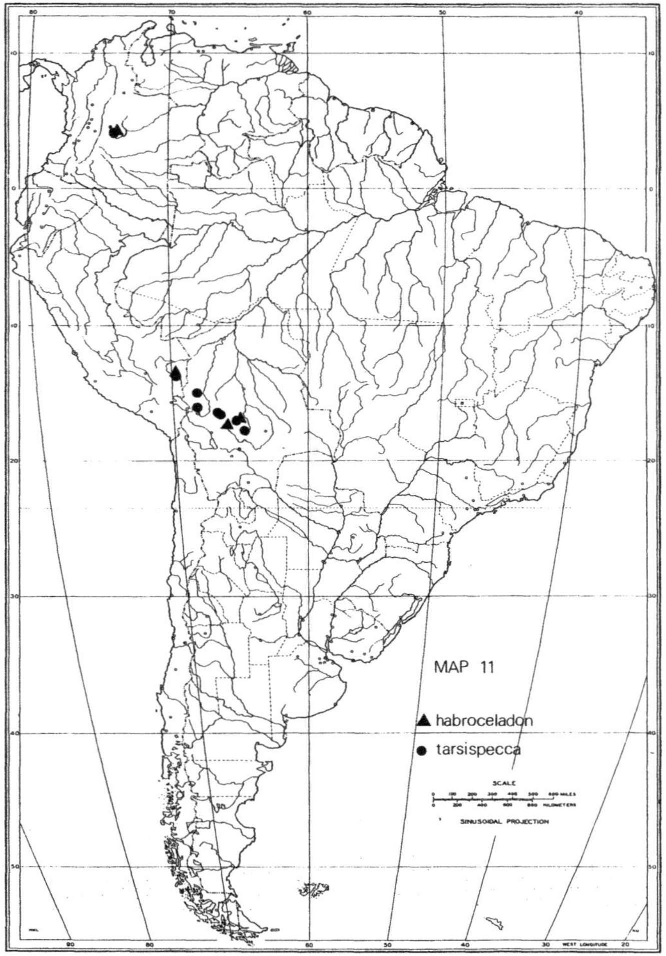

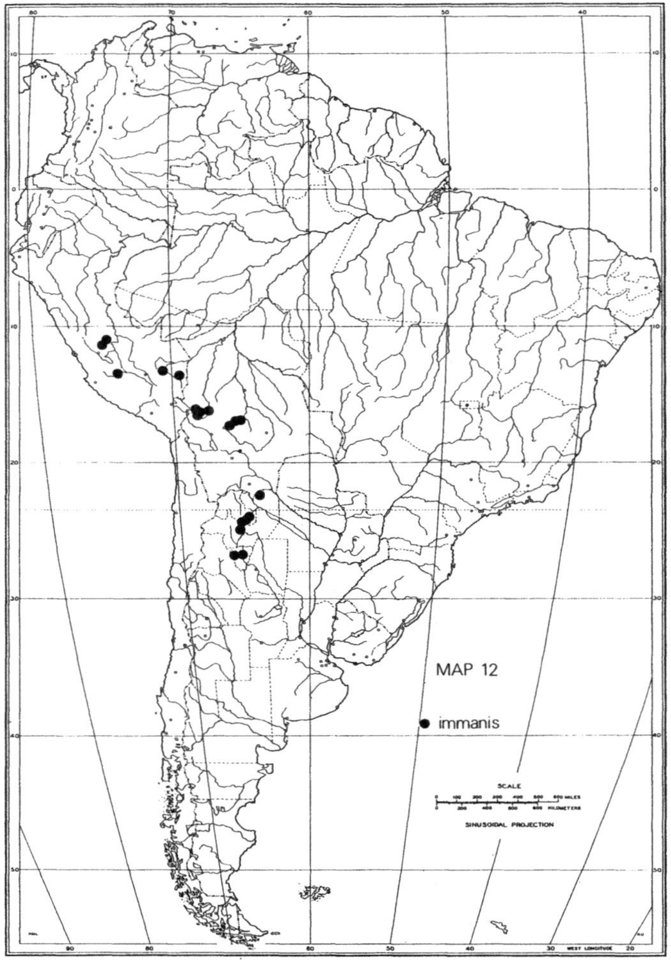

Map 12. *M. immanis*

Map 11. *M. habroceladon, M. tarsispecca*

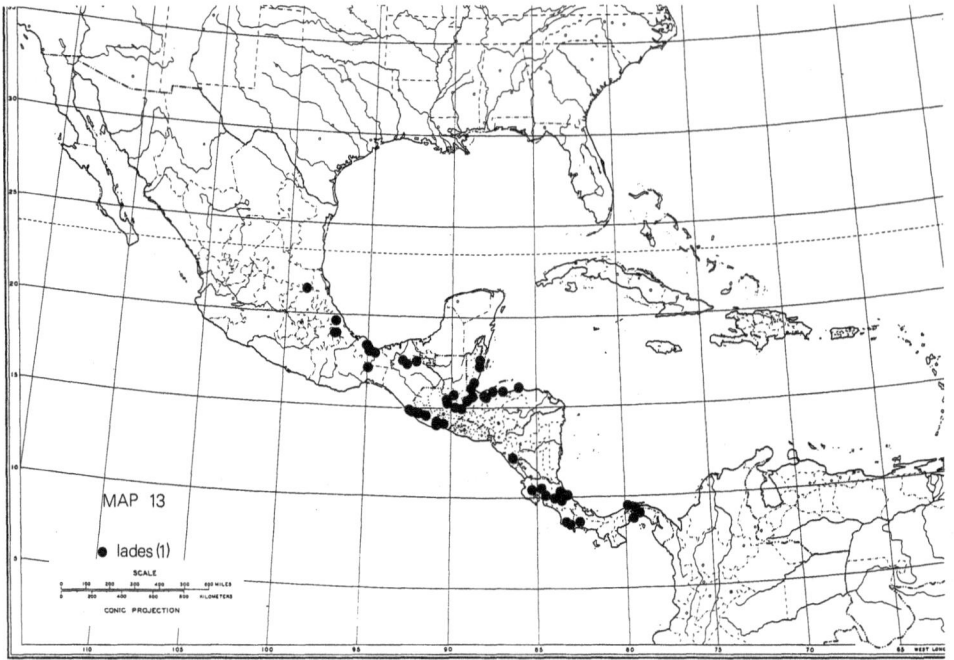

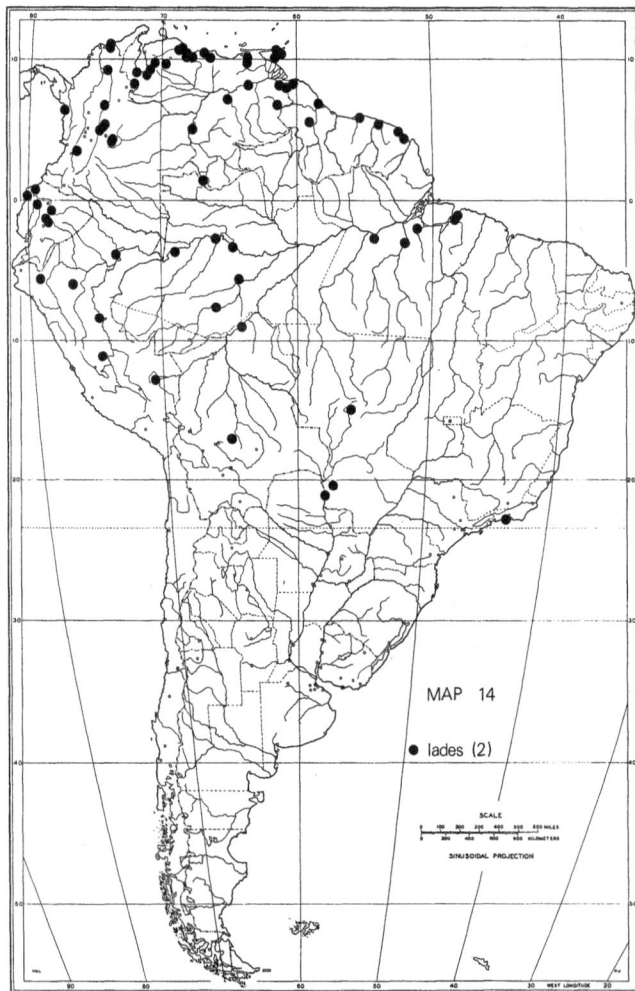

Maps 13 and 14. *M. lades*

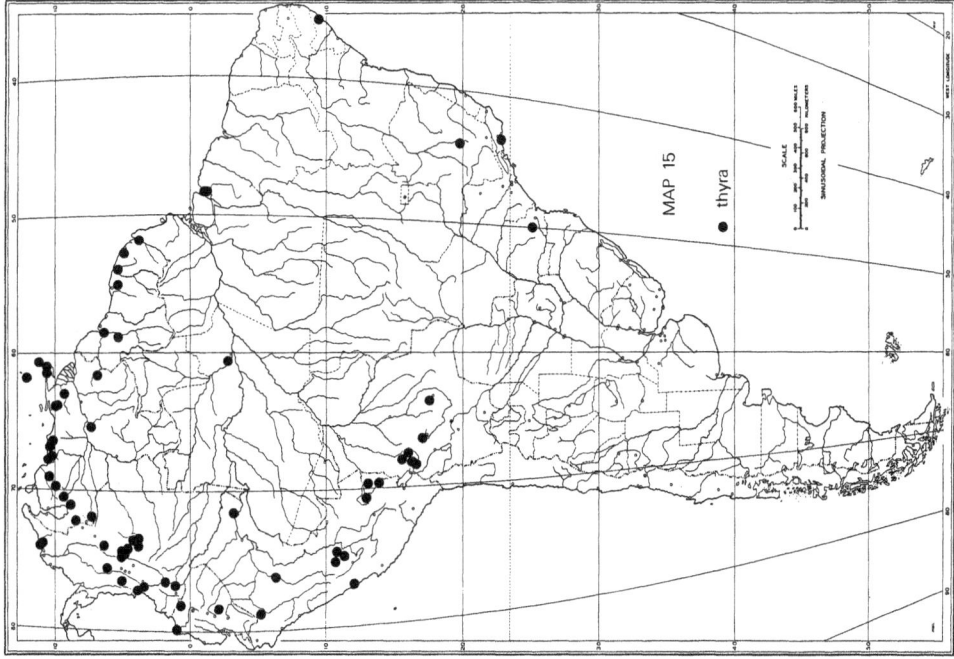

Map 16. *M. thyridia, M. pelotas*

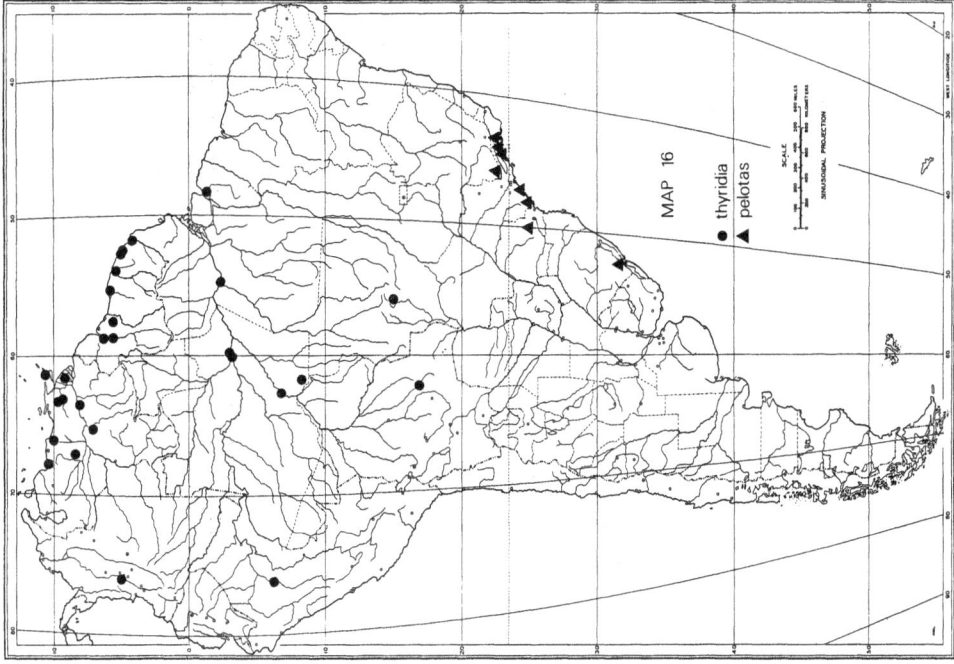

Map 15. *M. thyra*

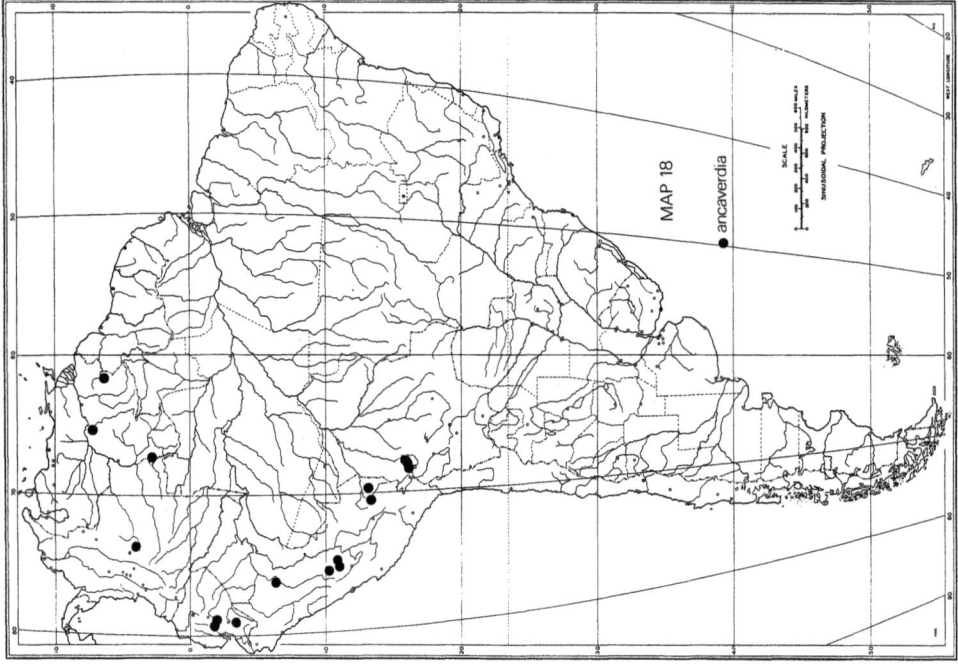

Map 18. *M. ancaverdia*

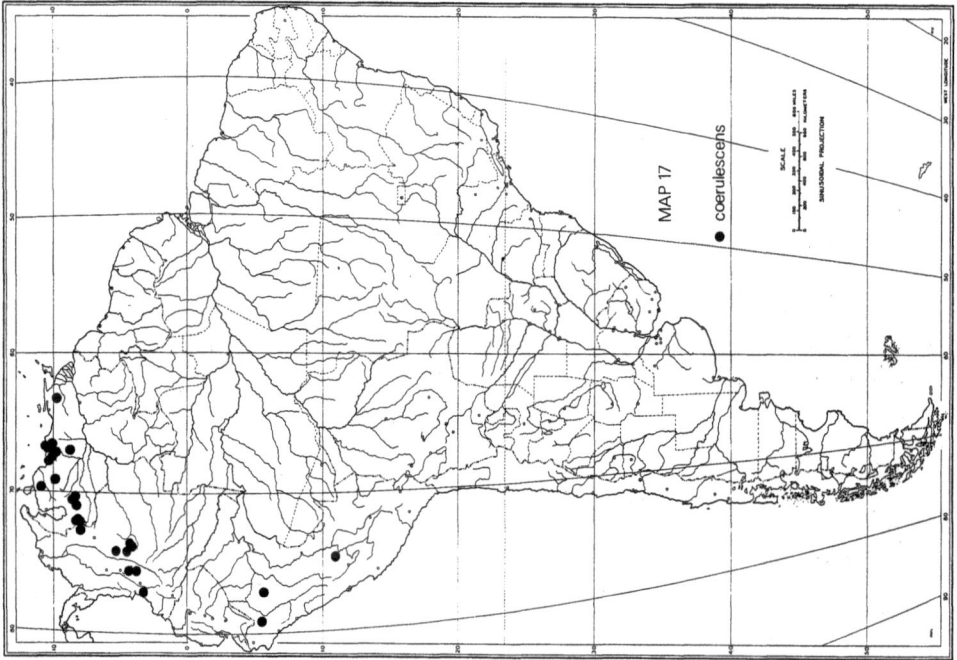

Map 17. *M. coerulescens*

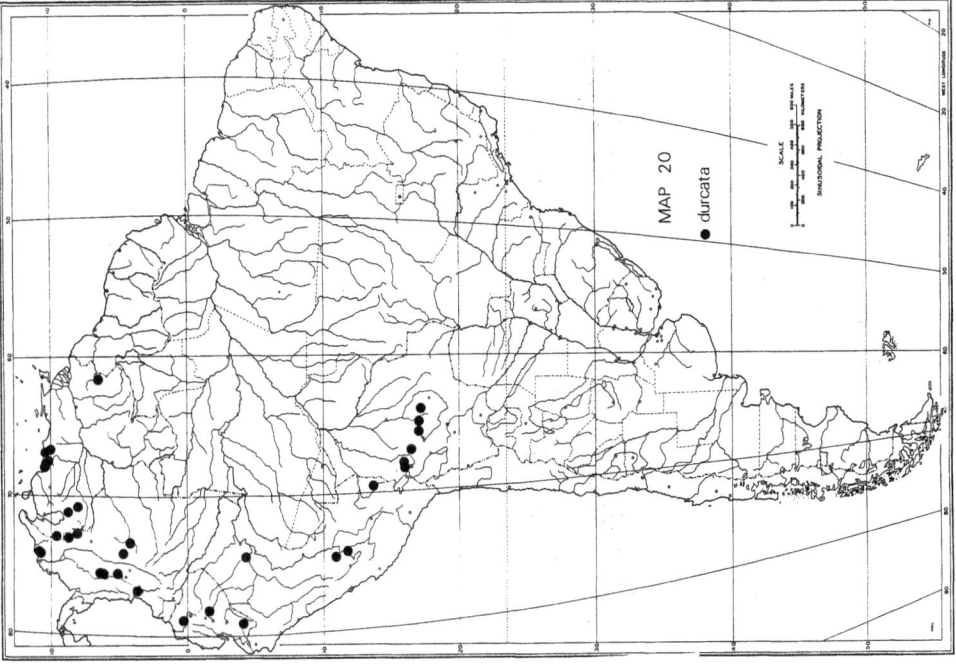

Map 20. *M. durcata*

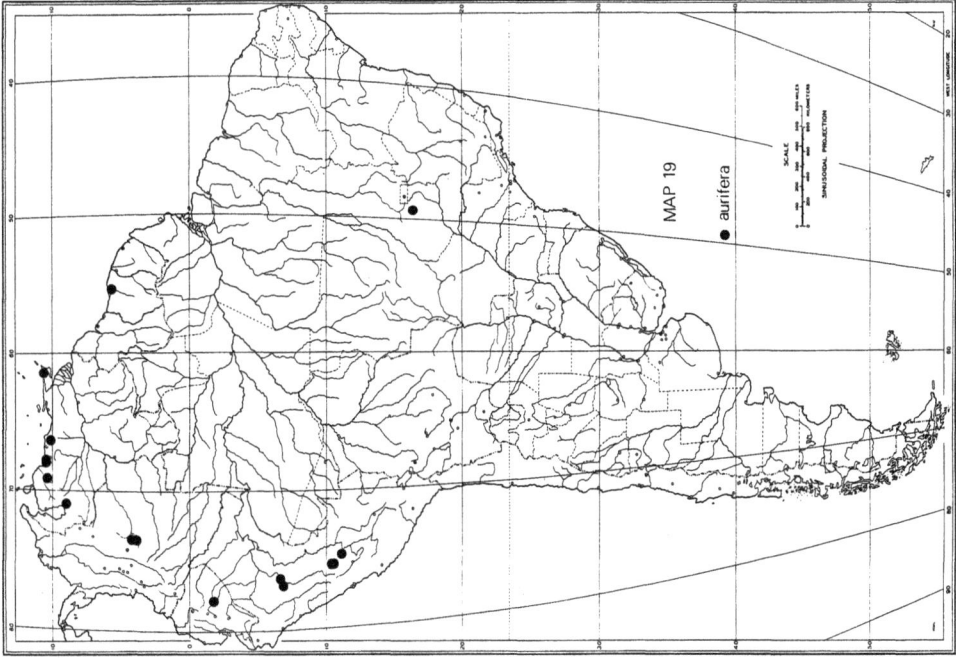

Map 19. *M. aurifera*

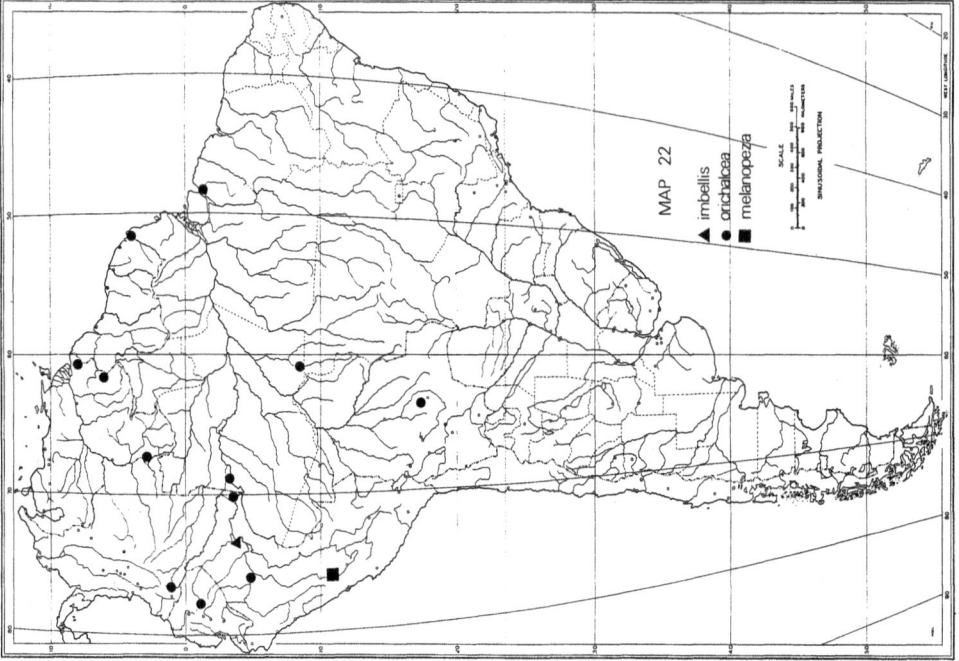

Map 22. *M. imbellis, M. orichalcea, M. melanopeza*

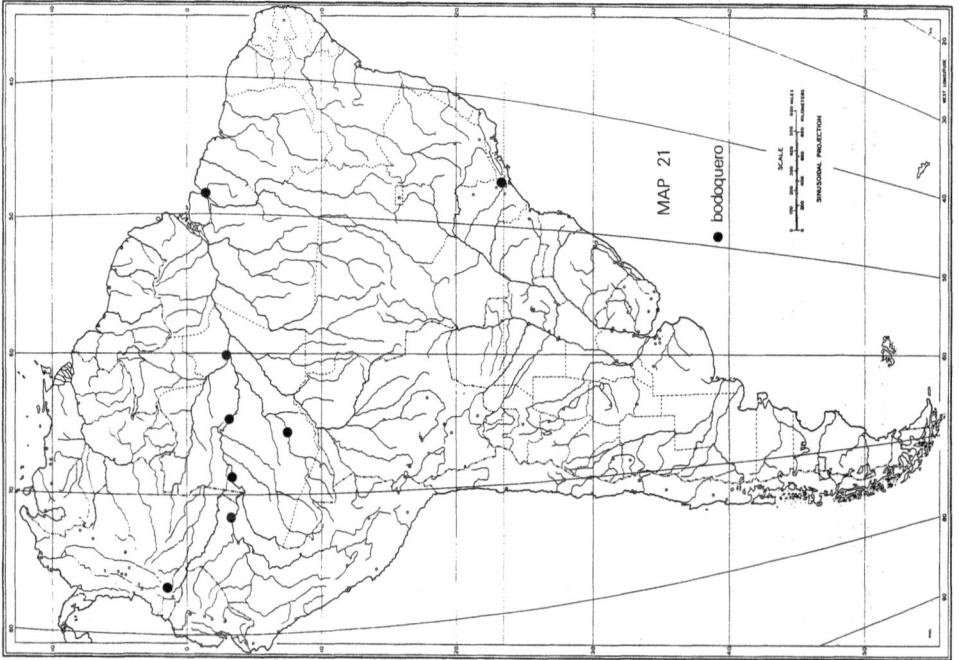

Map 21. *M. bodoquero*

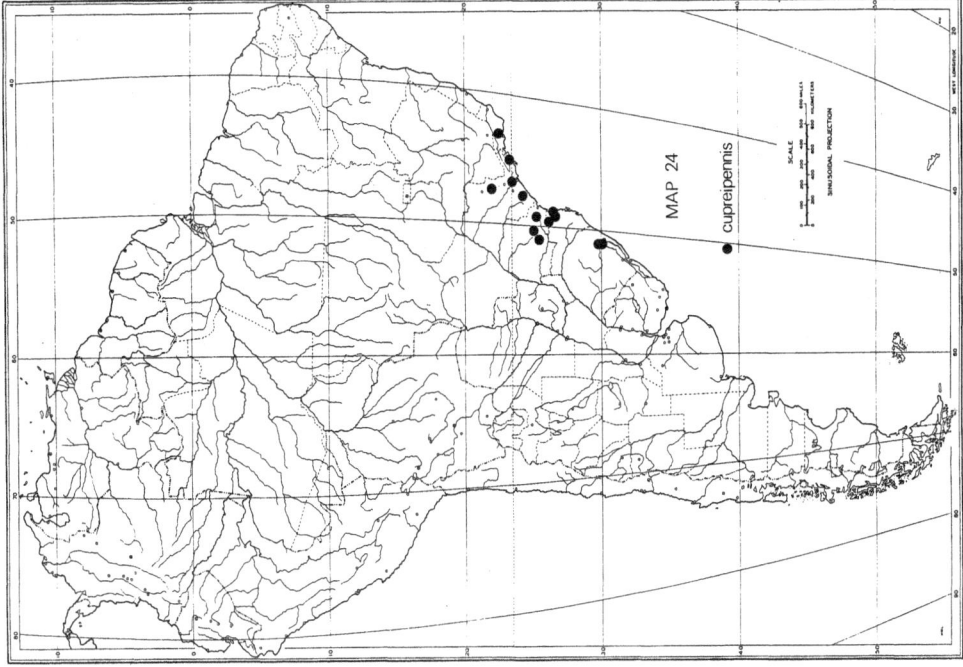

Map 24. *M. cupreipennis*

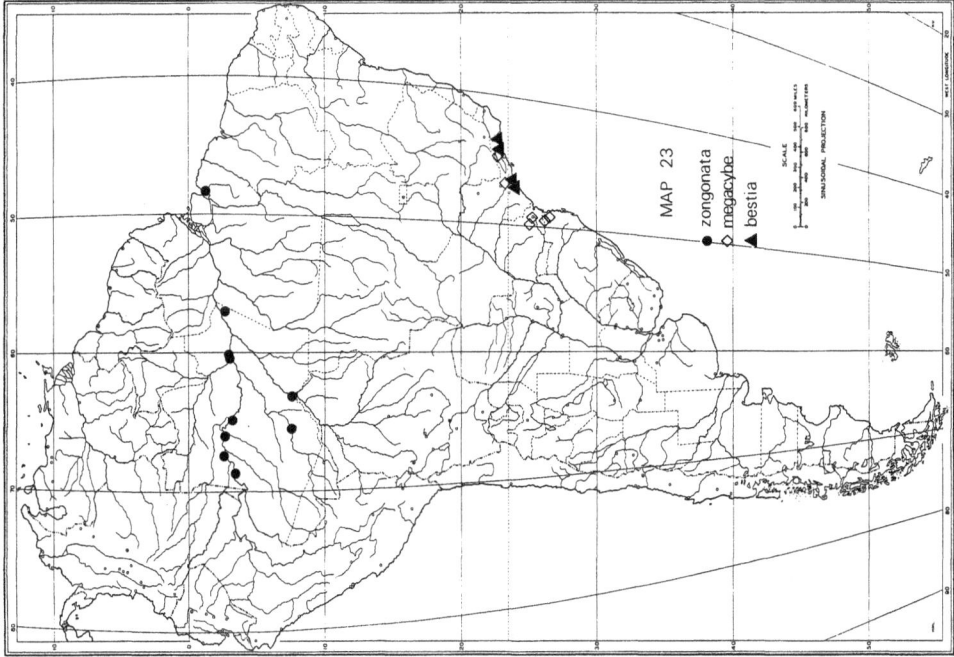

Map 23. *M. zongonata, M. megacybe, M. bestia*

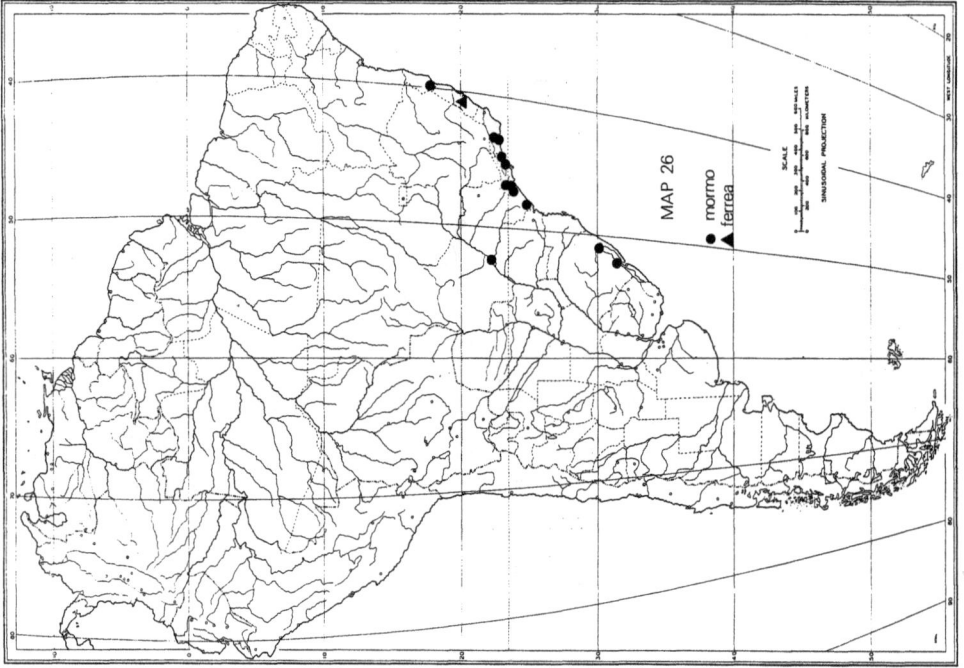

Map 26. *M. mormo, M. ferrea*

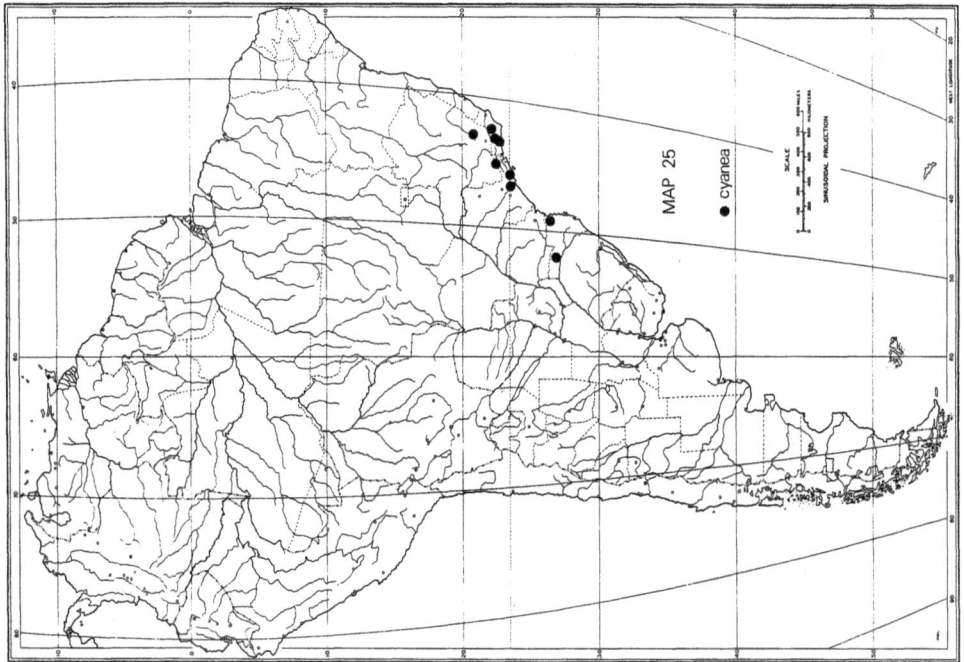

Map 25. *M. cyanea*

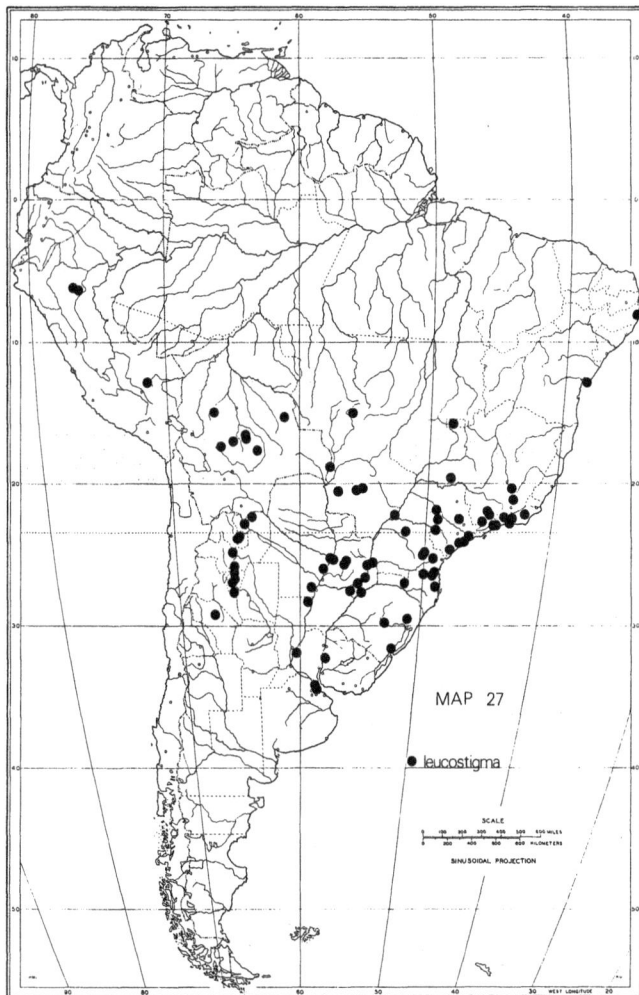

Map 27. *M. leucostigma*

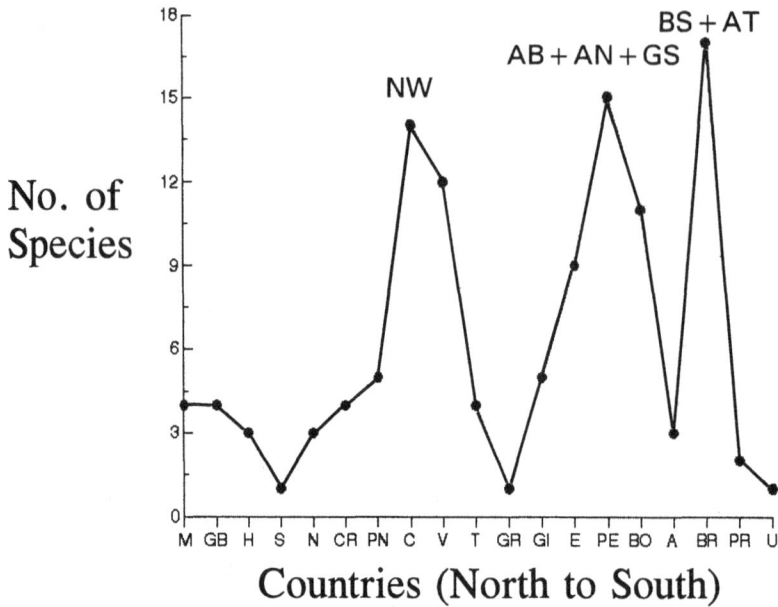

Figure 1. Distribution of *Macrocneme* species by country, with corresponding biogeographical regions from Map 1 designated for the major peak areas. M = Mexico; GB = Guatemala/Belize; H = Honduras; S = El Salvador; N = Nicaragua; CR = Costa Rica; PN = Panama; C = Colombia; V = Venezuela; T = Trinidad; GR = Grenada; GI = the Guianas; E = Ecuador; PE = Peru; BO = Bolivia; A = Argentina; BR = Brazil; PR = Paraguay; U = Uruguay; NW = Northwestern; AB+AN+GS = Amazonian Hyalea; BS+AT = Atlantic.

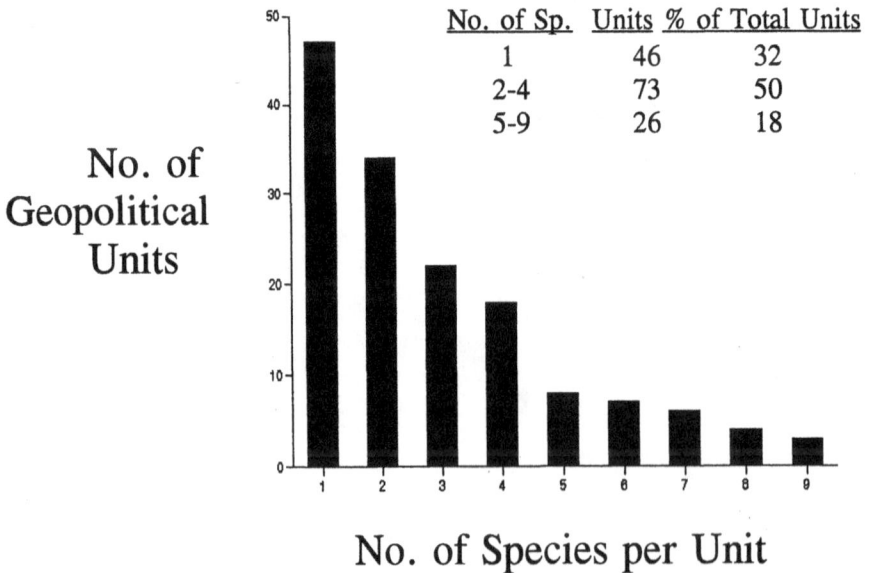

Figure 2. Comparative sympatry in *Macrocneme*, by political boundaries from Mexico to Argentina. Units represent political subdivisions as either states, provinces, or departments.

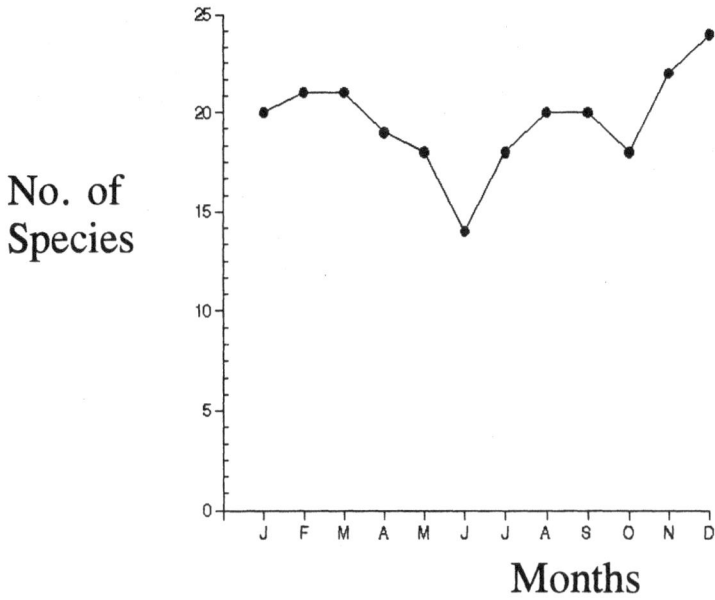

Figure 3. Seasonal availability for species of adult *Macrocneme*.

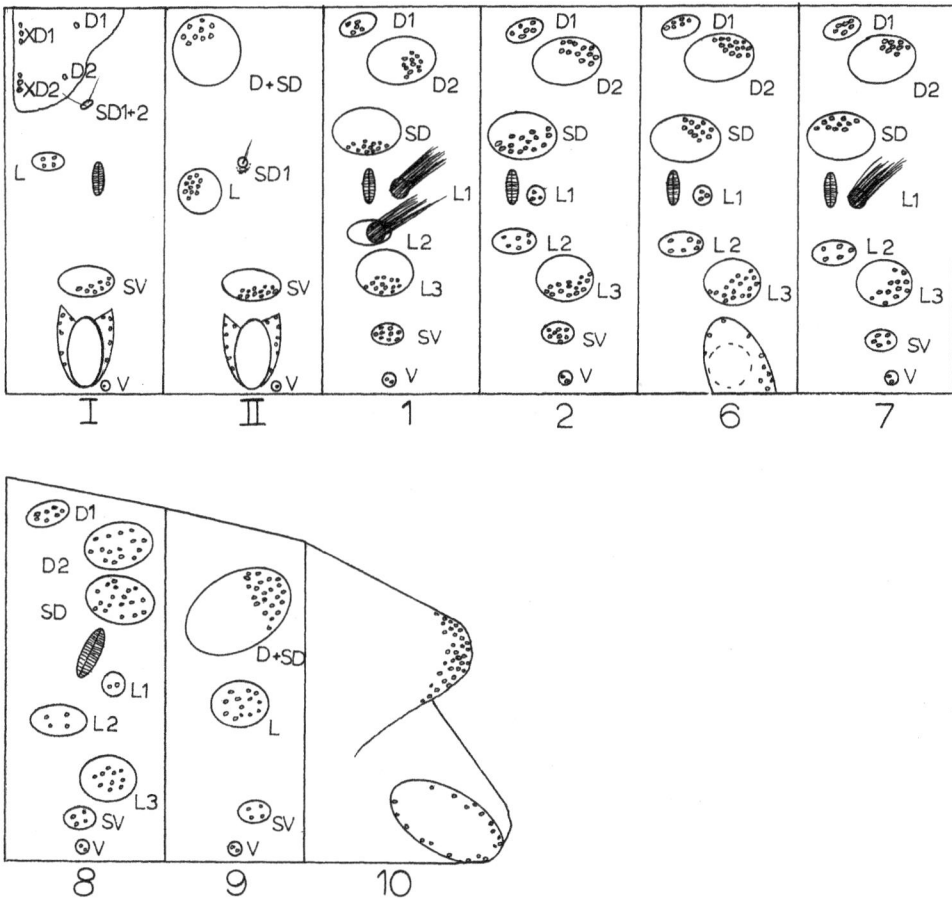

Figure 4. Larval chaetotaxy for species of *Macrocneme*: lateral view of prothorax, mesothorax, and abdominal segments 1, 2, and 6 through 10.

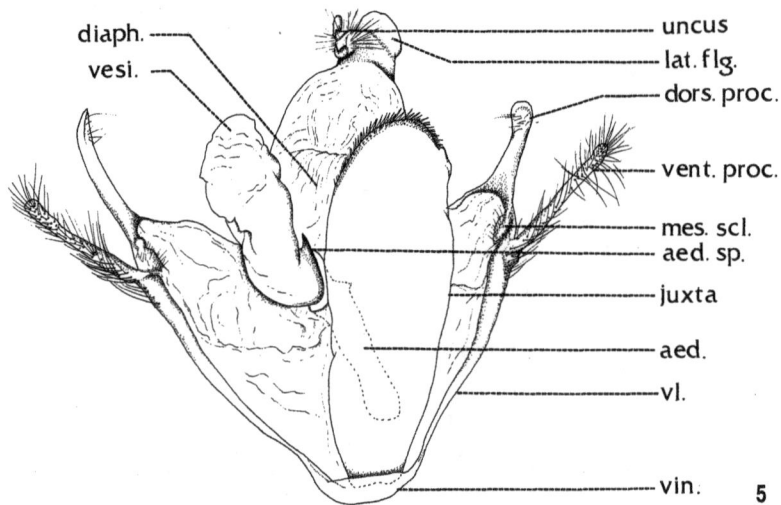

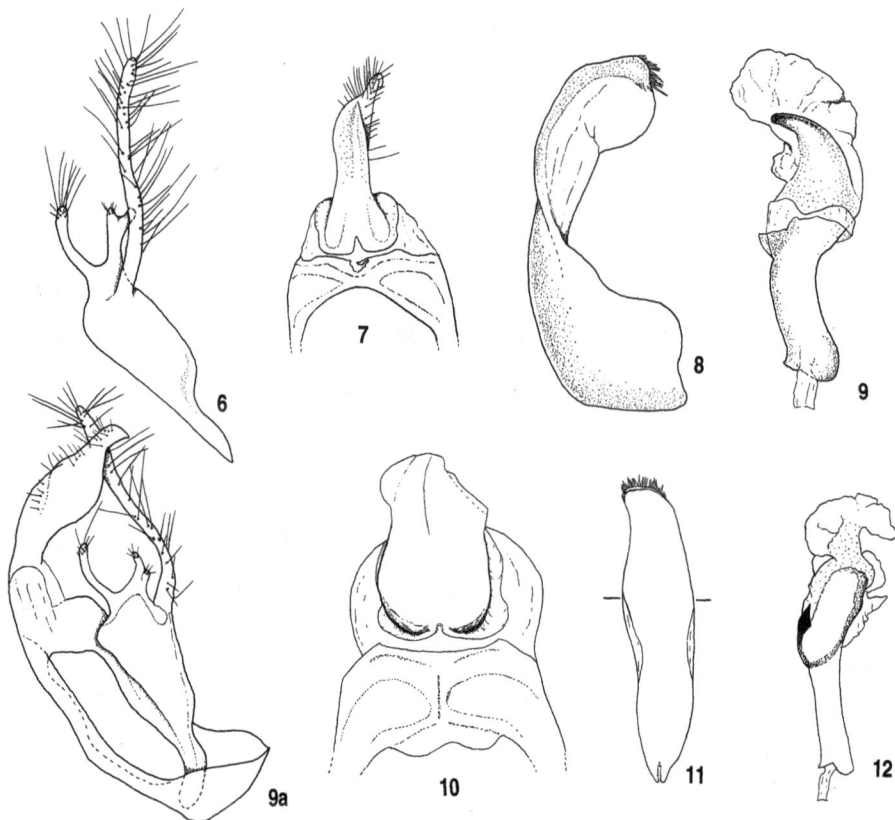

Figures **5-12.** Male genitalia. Fig. **5.** Ventral view showing parts *in situ* (from male holotype, *M. melanopeza*): lat. flg. = lateral flange of uncus; dors. proc. = dorsal process of valva; vent. proc. = ventral process of valva; mes. scl. = mesal sclerite of valva; aed. sp. = aedeagal spine; aed. = aedeagus; vl. = valva; vin. = vinculum; diaph. = diaphragma; vesi. = vesica. Figs. **6-9.** *M. adonis* (in four parts, l. to r.): left valva, uncus (dorsum), juxta, aedeagus; **9a**, latero-ventral view with aedeagus and left valva removed. Figs. **10-12.** *M. chrysitis* (l. to r.): uncus (dorsum), juxta, aedeagus (from *iole* syntype, San Gerónimo, Guatemala).

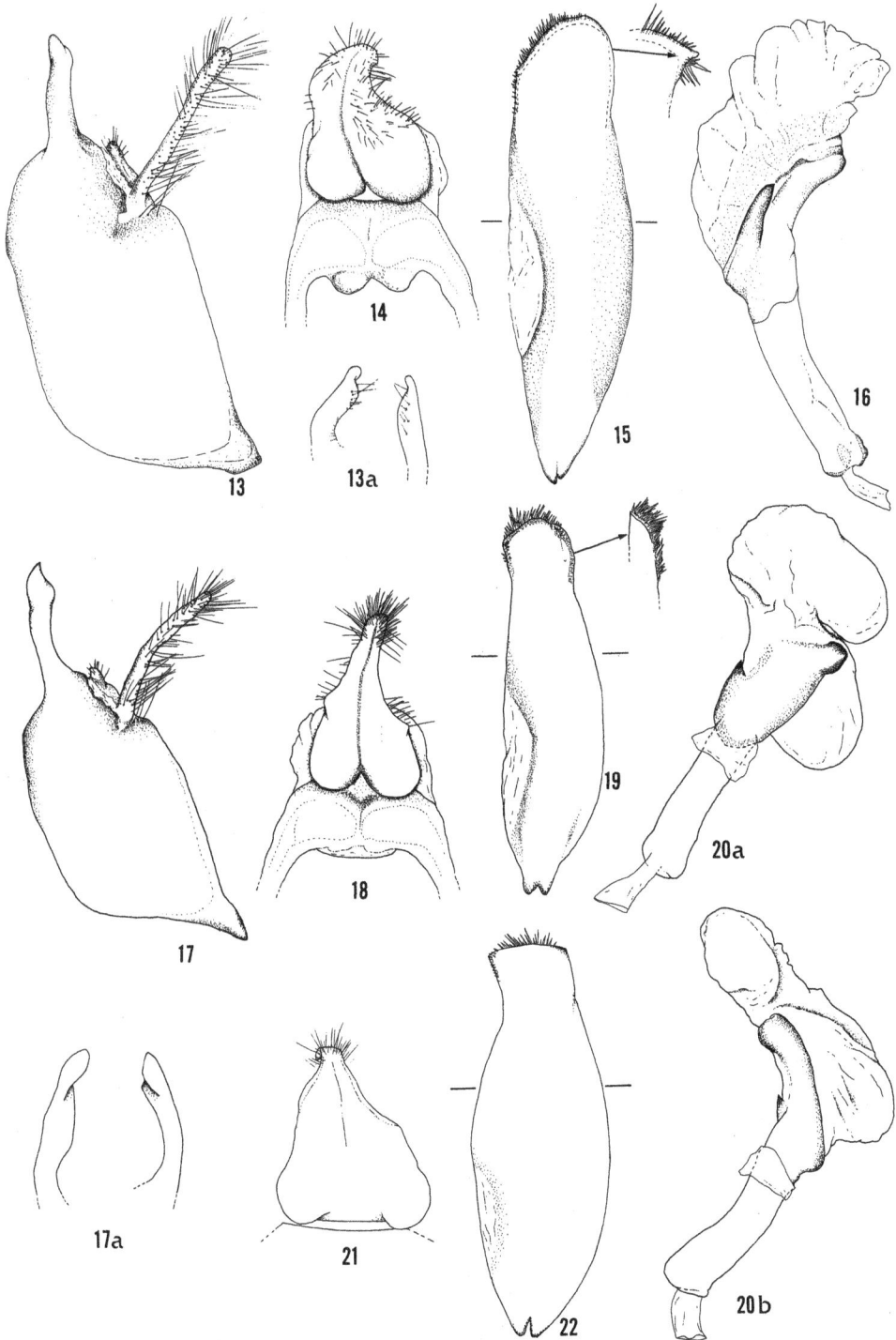

Figures 13-22. Male genitalia (in four parts, l. to r.): left valva, uncus (dorsum), juxta, aedeagus. Figs. 13-16. *M. chrysitis*, Jacala, Mexico; 13a, dorsal processes of valvae. Figs. 17-20. *M. iole*, Rovira, Panamá; 17a, dorsal processes of valvae, viewed dorsally to show inner margins; 20a-b, ventral and lateral views of aedeagus. Figs. 21-22. *M. iole*, Villa Somoza, Nicaragua, variations in uncus and juxta.

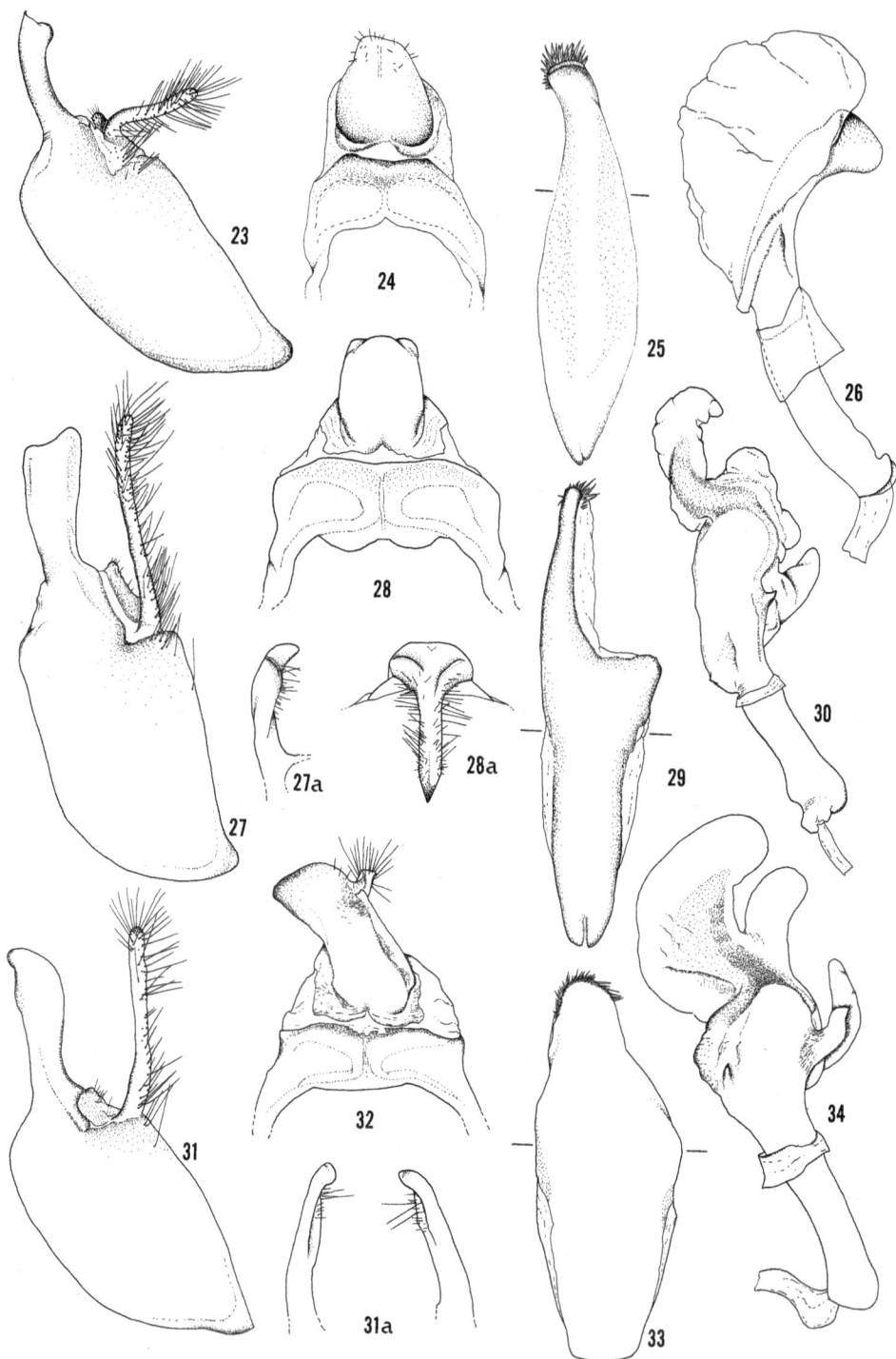

Figures **23-34.** Male genitalia (in four parts, l. to r.): left valva, uncus (dorsum), juxta, aedeagus. Figs. **23-26.** *M. semiviridis*, Rancho Grande, Venezuela. Figs. **27-30.** *M. cabimensis*, Colombia; **27a,** apex of left dorsal process showing concave inner margin; **28a,** dorsal view of uncus. Figs. **31-34.** *M. oponiensis*, Alto Rio Opón, Colombia; **31a,** dorsal processes of valvae, viewed dorsally to show inner margins.

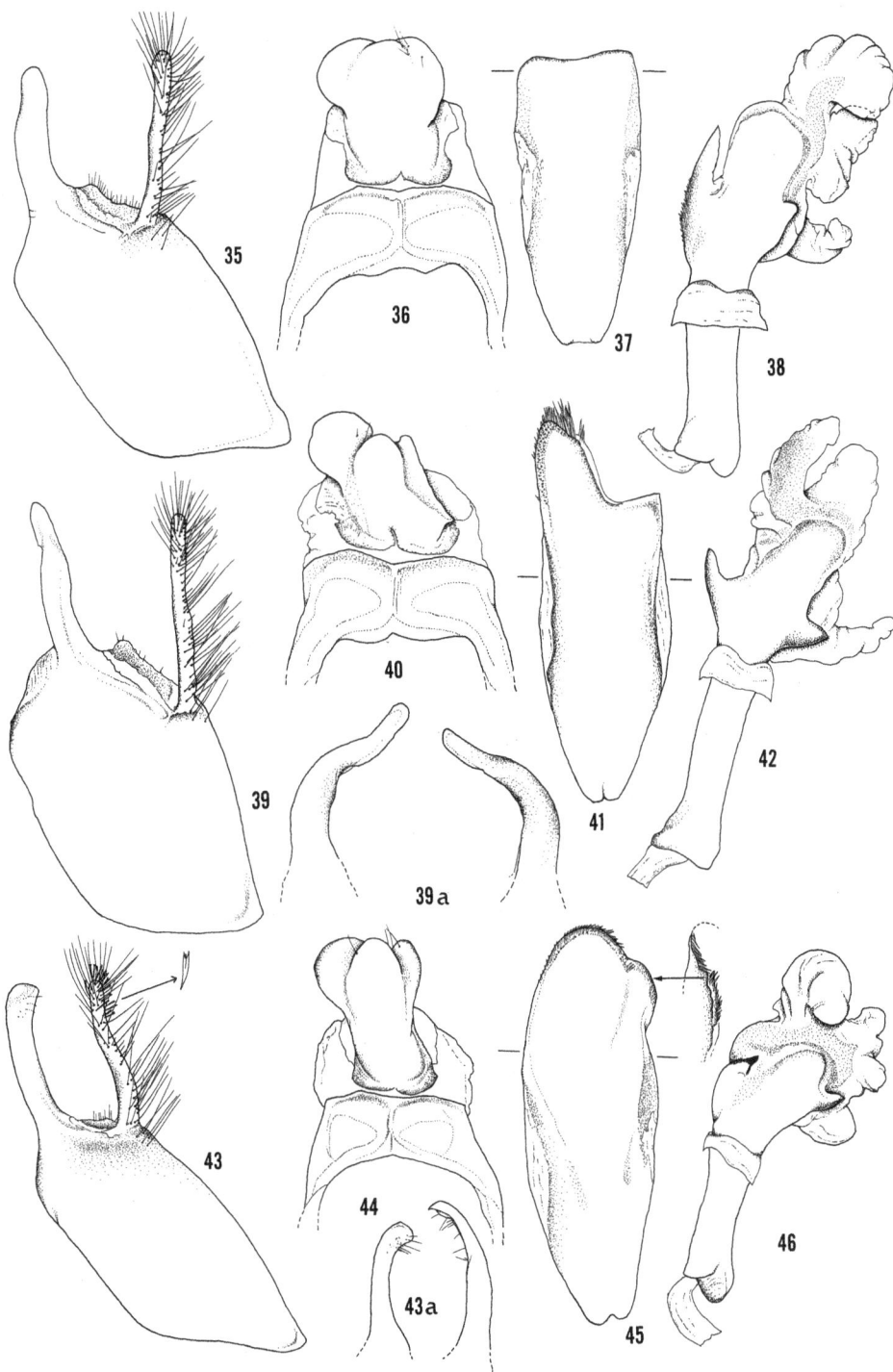

Figures 35-46. Male genitalia (in four parts, l. to r.): left valva, uncus (dorsum), juxta, aedeagus. Figs. 35-38. *M. immanis*, Yungas del Palmar, Bolivia. Figs. 39-42. *M. tarsispecca*, Yungas del Palmar, Bolivia; 39a, dorsal processes of valvae, viewed dorsally to show curvature of tips. Figs. 43-46. *M. habroceladon*, Yungas del Palmar, Bolivia; 43a, dorsal processes of valvae, viewed dorsally to show inner margins.

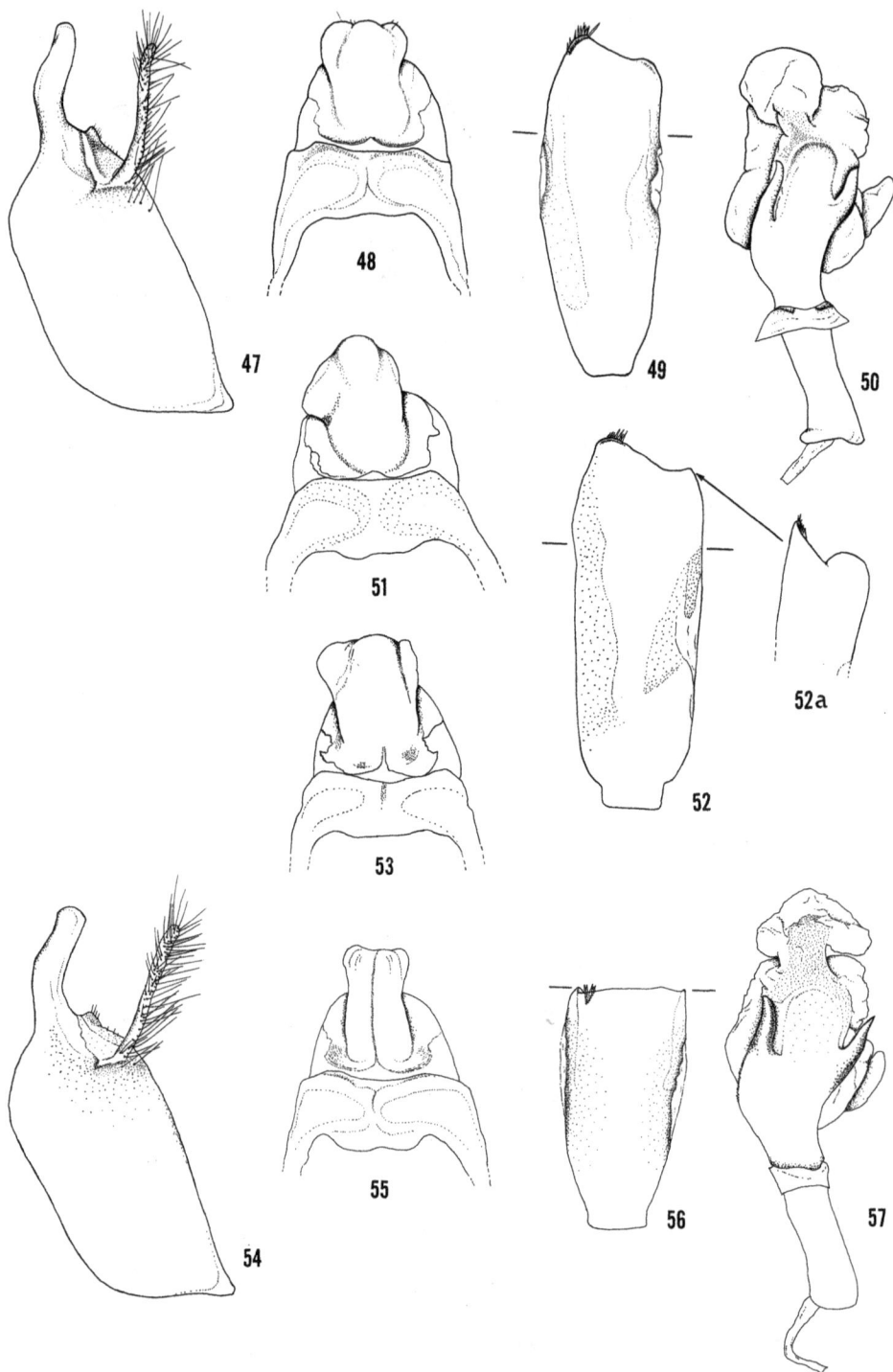

Figures 47-57. Male genitalia (in four parts, l. to r.): left valva, uncus (dorsum), juxta, aedeagus. Figs. 47-50. *M. lades*, Choroní, Venezuela. Figs. 51-53. *M. lades*, variations in uncus and juxta: 51-52a, Zamora, Ecuador; 53, Río Morichal Largo, Venezuela, variation in uncus. Figs. 54-57. *M. thyra*, Choroní, Venezuela.

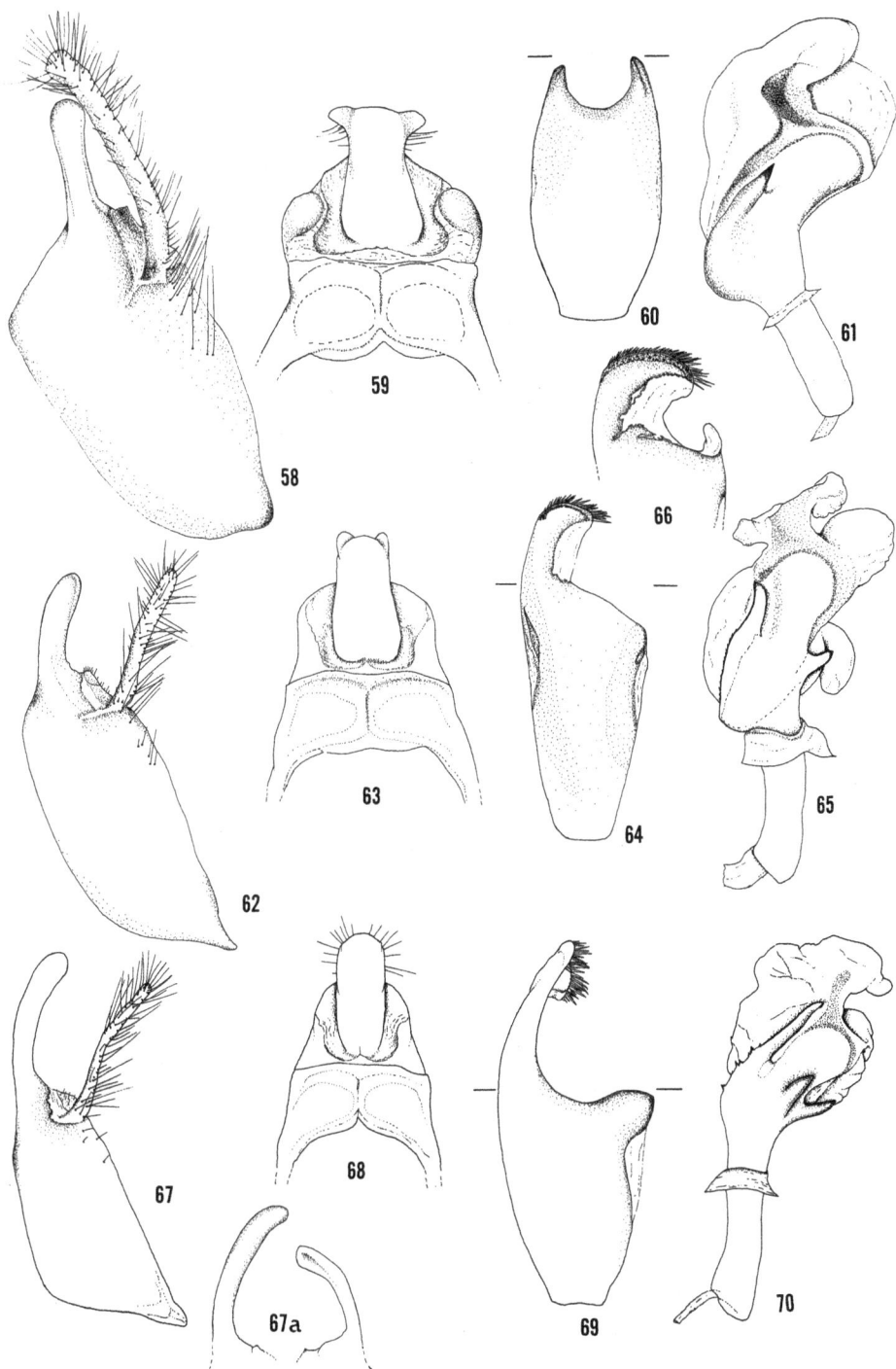

Figures 58-70. Male genitalia (in four parts, l. to r.): left valva, uncus (dorsum), juxta, aedeagus. Figs. 58-61. *M. durcata*, Rancho Grande, Venezuela. Figs. 62-65. *M. coerulescens*, San Isidro, Barinas, Venezuela. Fig. 66. *M. coerulescens*, Barinitas, Venezuela, variation in r. margin of juxta. Figs. 67-70. *M. thyridia*, Rio Morichal Largo, Venezuela; 67a, dorsal processes of valvae, viewed dorsally to show asymmetry.

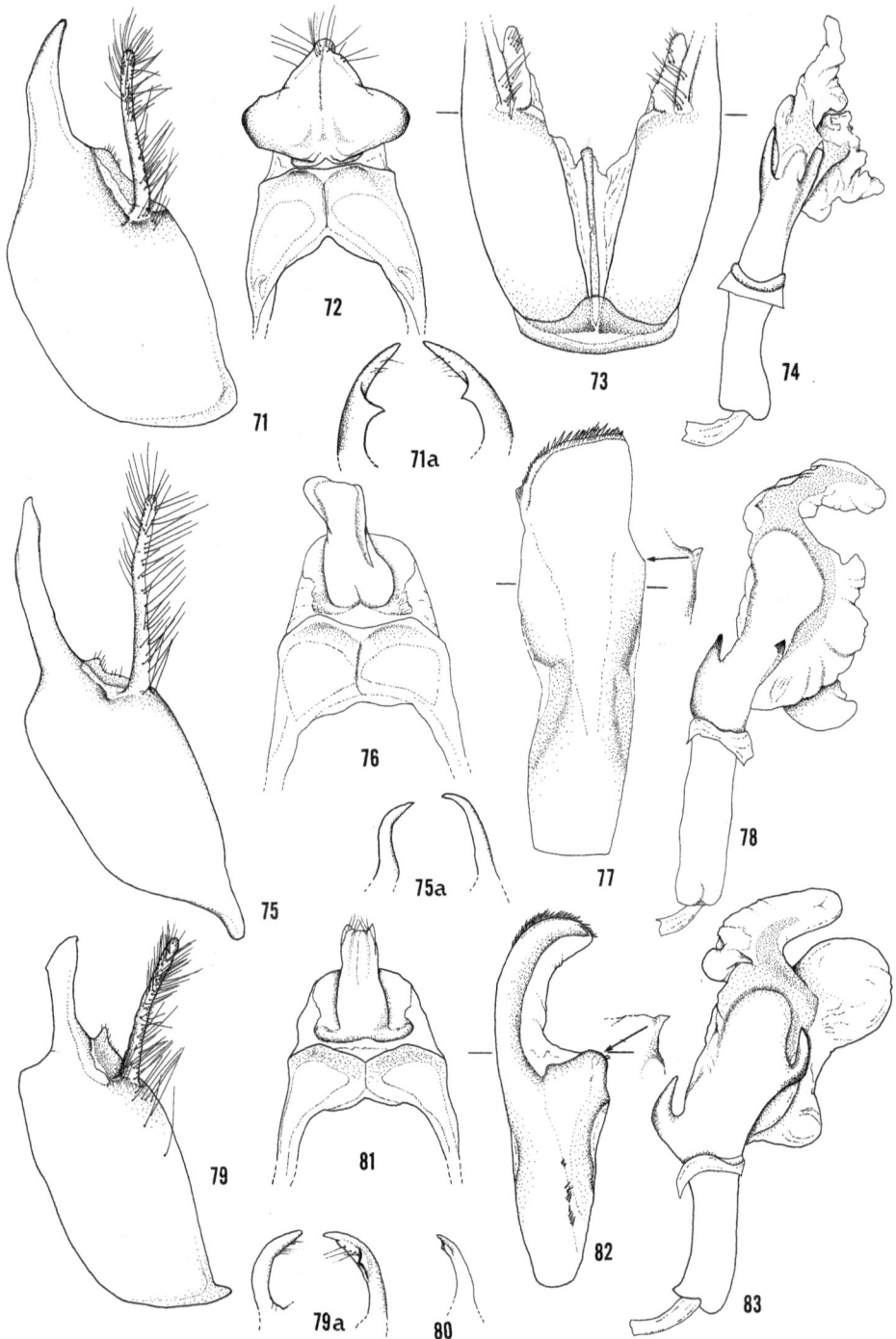

Figures **71-83.** Male genitalia (in four parts, l. to r.): left valva, uncus (dorsum), juxta, aedeagus. Figs. **71-74.** *M. aurifera*, Guatopo National Park, Venezuela; **71a,** dorsal process of valvae, viewed dorsally to show spined margins. Figs. **75-78.** *M. ancaverdia*, Yavita, Venezuela; **75a** dorsal processes of valvae, viewed dorsally to show asymmetry (not to same scale as Fig. 75). Figs. **79-83.** *M. orichalcea*, Bolívar, Venezuela; **79a,** dorsal processes of valvae, viewed dorsally to show spine on left arm; **80,** variation of l. arm of dorsal process of valva, (ex Perú).

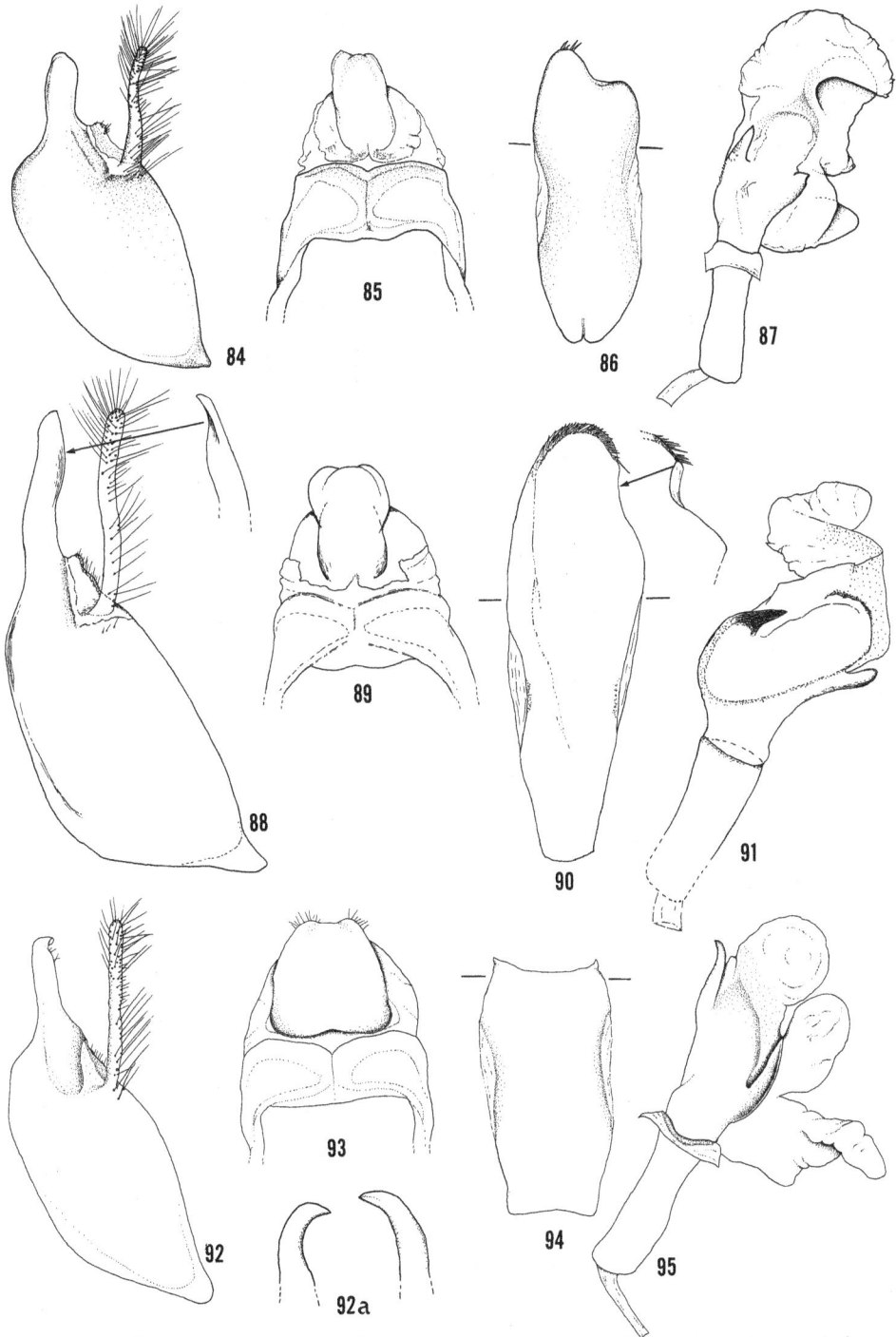

Figures **84-95.** Male genitalia (in four parts, l. to r.): left valva, uncus (dorsum), juxta, aedeagus. Figs. **84-87.** *M. bodoquero*, Caquetá, Colombia. Figs. **88-91.** *M. imbellis*, holotype, Iquitos, Perú. Figs. **92-95.** *M. zongonata*, Oroya, Perú; **92a,** dorsal processes, viewed dorsally to show asymmetry of arms.

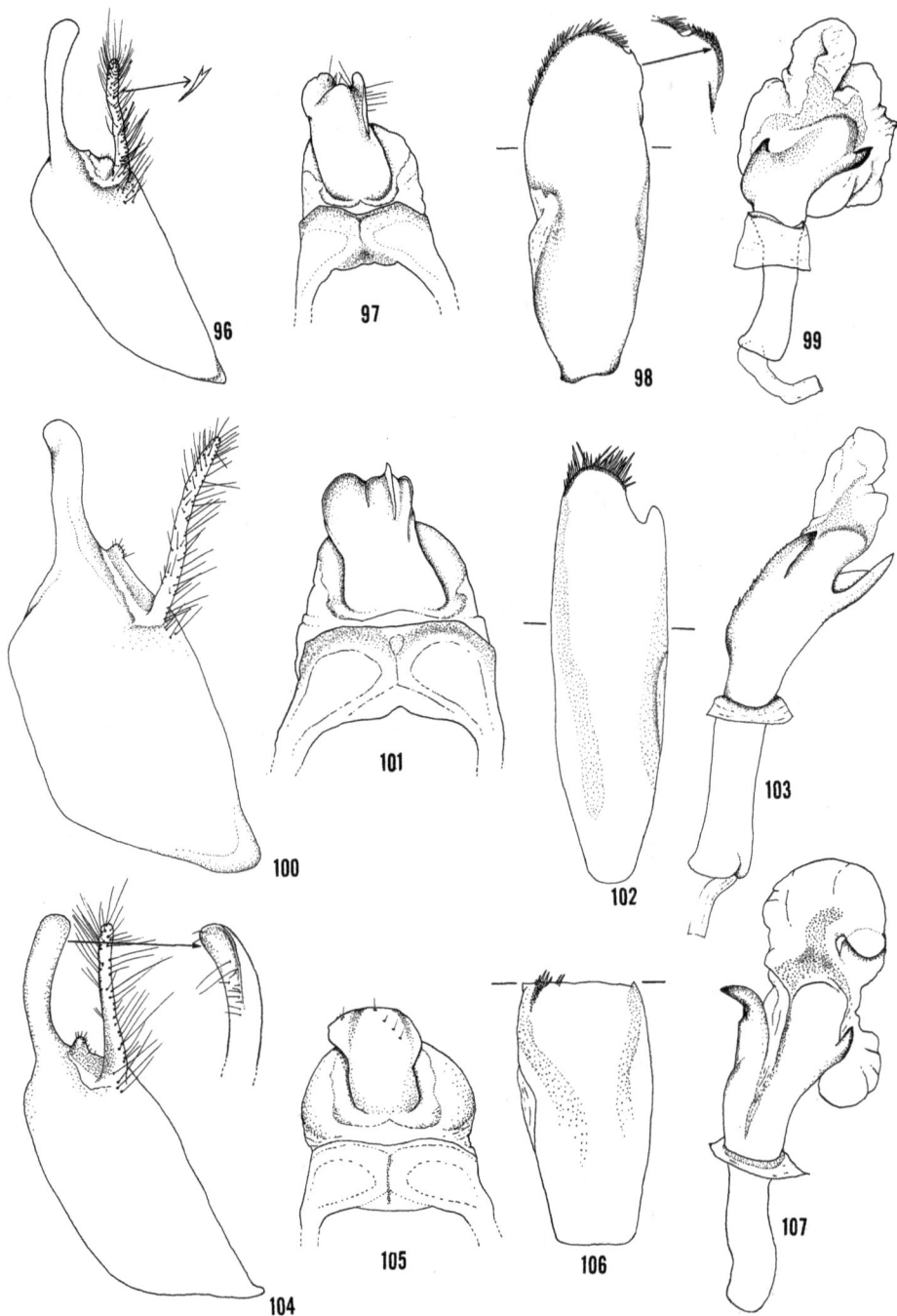

Figures **96-107.** Male genitalia (in four parts, l. to r.): left valva, uncus (dorsum), juxta, aedeagus. Figs. **96-99.** *M. melanopeza*, holotype, Perú. Figs. **100-103.** *M. leucostigma*, Asunción, Paraguay. Figs. **104-107.** *M. bestia*, Xerém, Brazil.

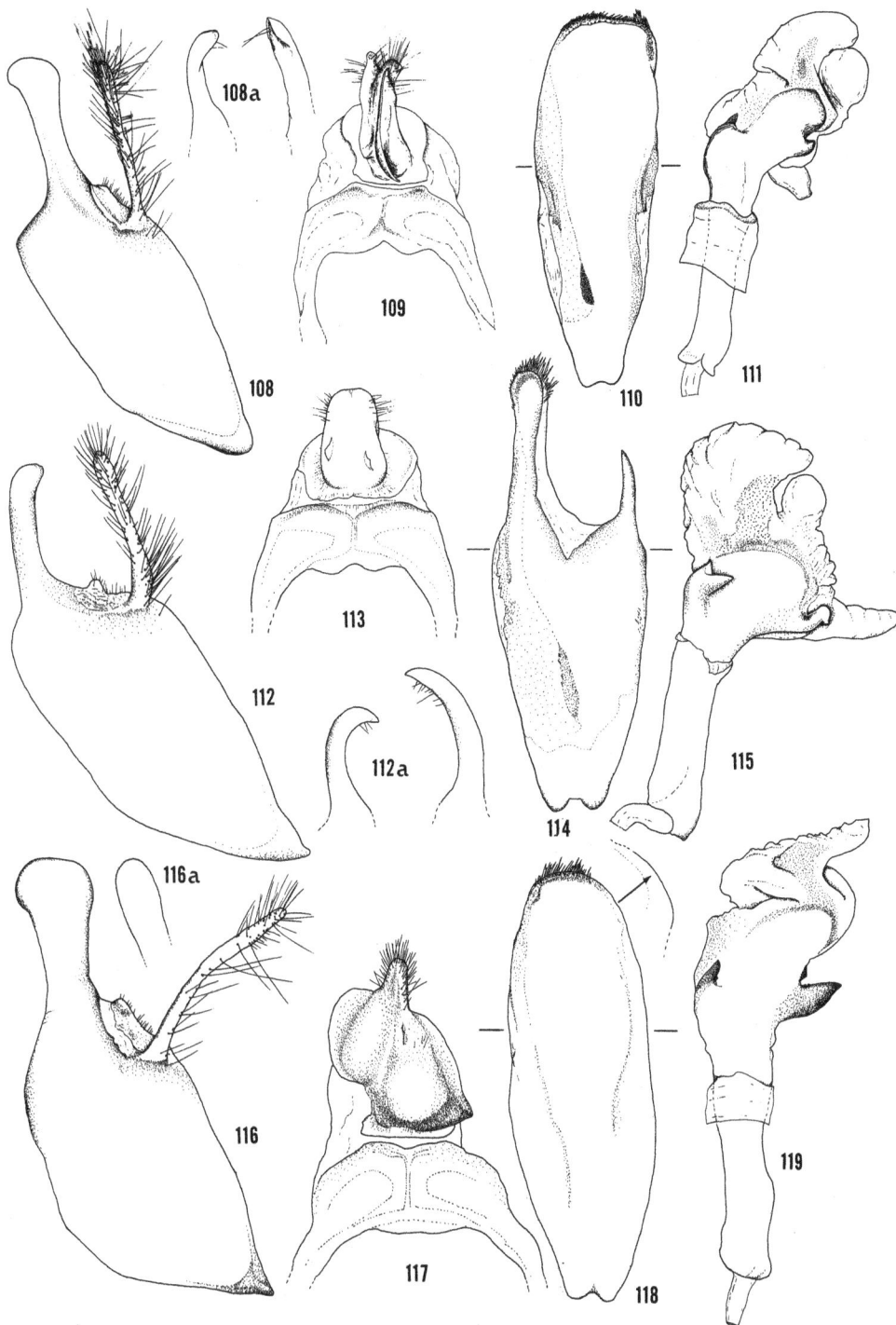

Figures **108-119.** Male genitalia (in four parts, l. to r.): left valva, uncus (dorsum), juxta, aedeagus. Figs. **108-111.** *M. cyanea*, Sta. Catarina, Brazil; **108a,** dorsal processes, viewed dorsally to show asymmetry of arms. Figs. **112-115.** *M. pelotas*, Pelotas, Brazil; **112a,** dorsal processes, viewed dorsally to show asymmetry of arms. Figs. **116-119.** *M. cupreipennis*, S. Paulo, Brazil; **116a,** outline of dorsal process of r. valva.

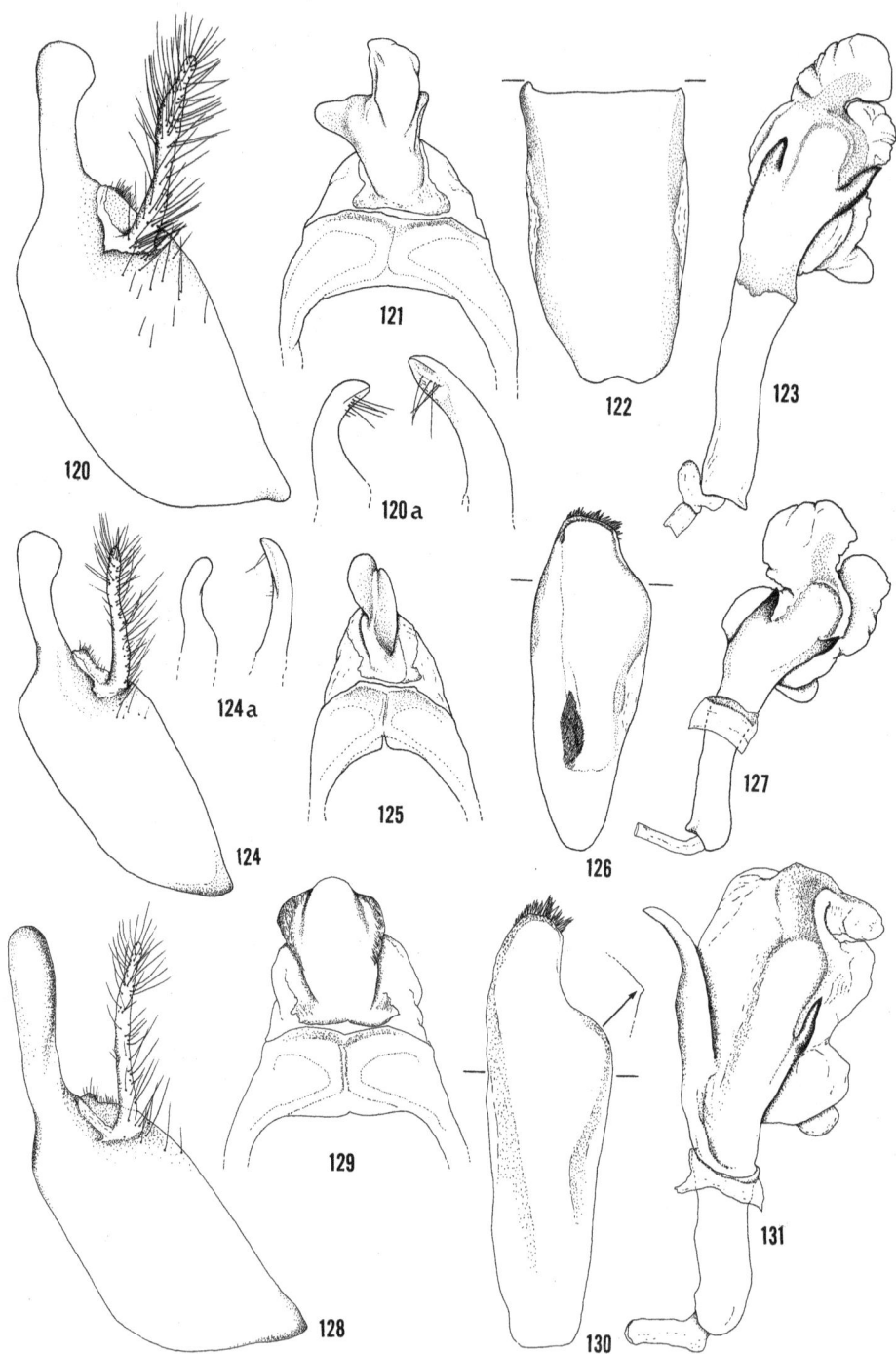

Figures 120-131. Male genitalia (in four parts, l. to r.): left valva, uncus (dorsum), juxta, aedeagus. Figs. 120-123. *M. megacybe*, Paraná, Brazil; 120a, dorsal processes, viewed dorsally to show asymmetry of arms. Figs. 124-127. *M. ferrea*, Sta. Teresa, Brazil; 124a, dorsal processes, viewed dorsally to show asymmetry of arms. Figs. 128-131. *M. mormo*, Rio Mucuri, Brazil.

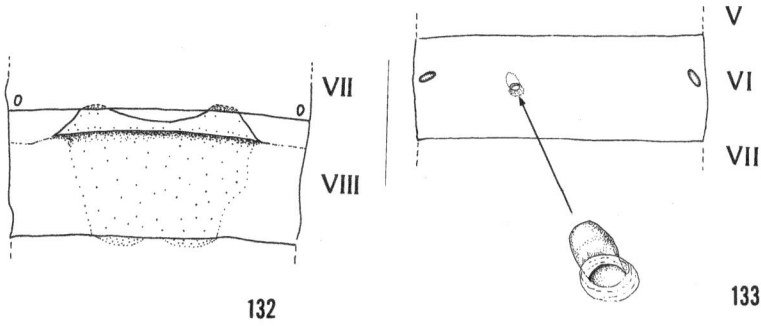

132   133

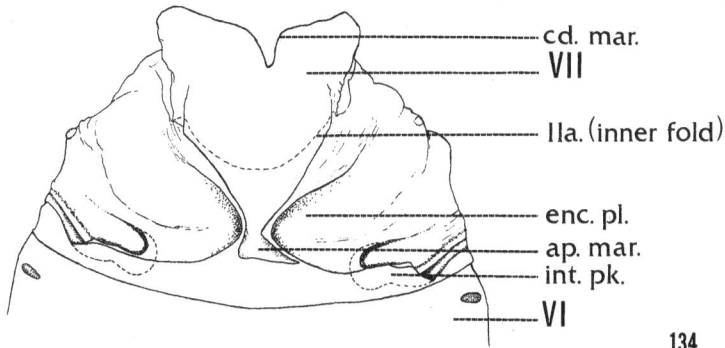

- cd. mar.
VII
- lla. (inner fold)
- enc. pl.
- ap. mar.
- int. pk.
VI
                                                              134

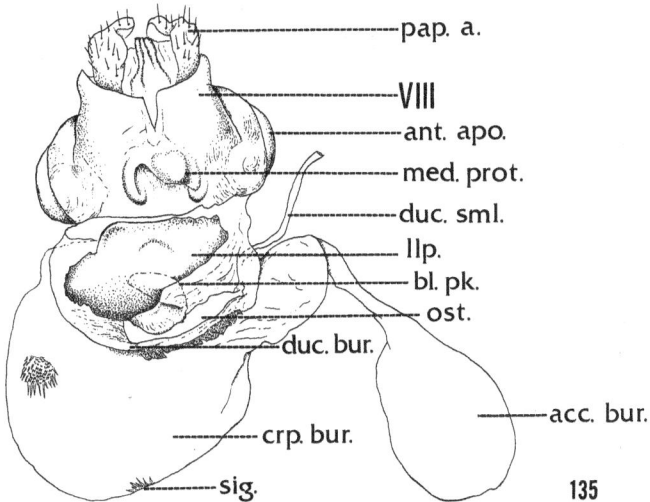

- pap. a.
VIII
- ant. apo.
- med. prot.
- duc. sml.
- llp.
- bl. pk.
- ost.
- duc. bur.
- acc. bur.
- crp. bur.
- sig.   135

Figure **132.** Intersegmental pocket between abdominal sternites VII and VIII in males of *Macrocneme*. Figure **133.** Pocket in abdominal sternite VI of male *tarsispecca*. Figure **134.** Venter of abdominal sternite VII in females of *M. durcata* showing descriptive features: cd. mar. = caudal margin; lla. = lamella antevaginalis; enc. pl. = encircling pleurite; ap. mar. = apical margin; int. pk. = intersegmental pocket. Figure **135.** Female genitalia (from *M. thyridia*), showing descriptive features: pap. a. = papillae anales; ant. apo. = anterior apophyses; med. prot. = medial protuberance of sternite VIII; duc. sml. = ductus seminalis; llp. = lamella postvaginalis; bl. pk. = blind pocket; ost. = ostium; duc. bur. = ductus bursae; crp. bur. = corpus bursae; sig. = signum; acc. bur. = accessory bursa.

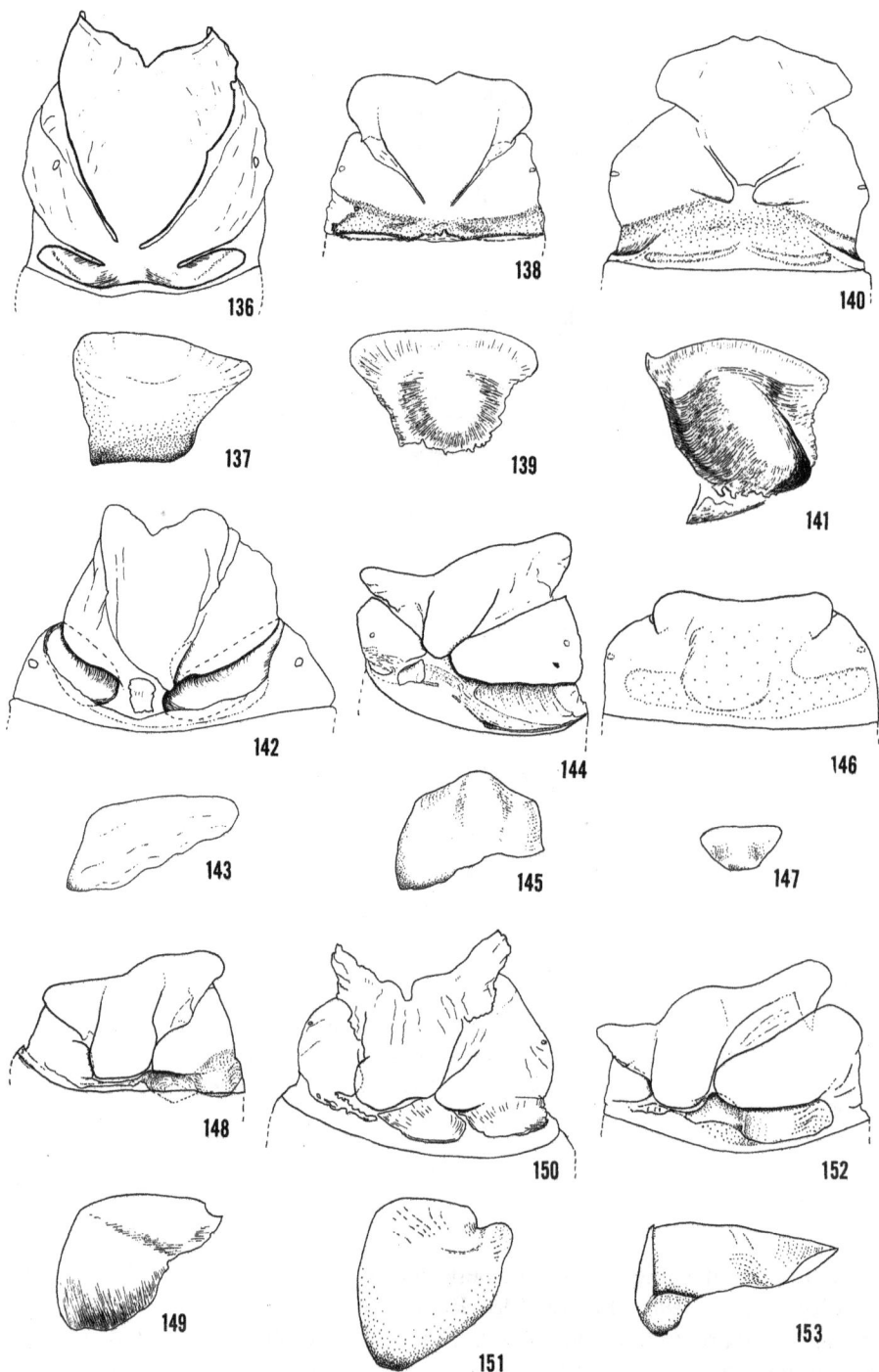

Figures 136-153. Abdominal sternite VII and sclerite of lamella postvaginalis in females. Figs. 136-137. *M. chrysitis*, Jacala, Mexico; 138-139. *M. iole*, San Isidro del General, Costa Rica; 140-141. *M. semiviridis*, Monterredondo, Colombia; 142-143. *M. cabimensis*, Colón, Panamá; 144-145. *M. oponiensis*, Villavicencio, Colombia; 146-147. *M. adonis*, Rancho Grande, Venezuela; 148-149. *M. lades*, Cachicamo, Venezuela; 150-151. *M. thyra*, Guatopo Nat. Park, Venezuela; 152-153. *M. ancaverdia*, La Merced, Perú.

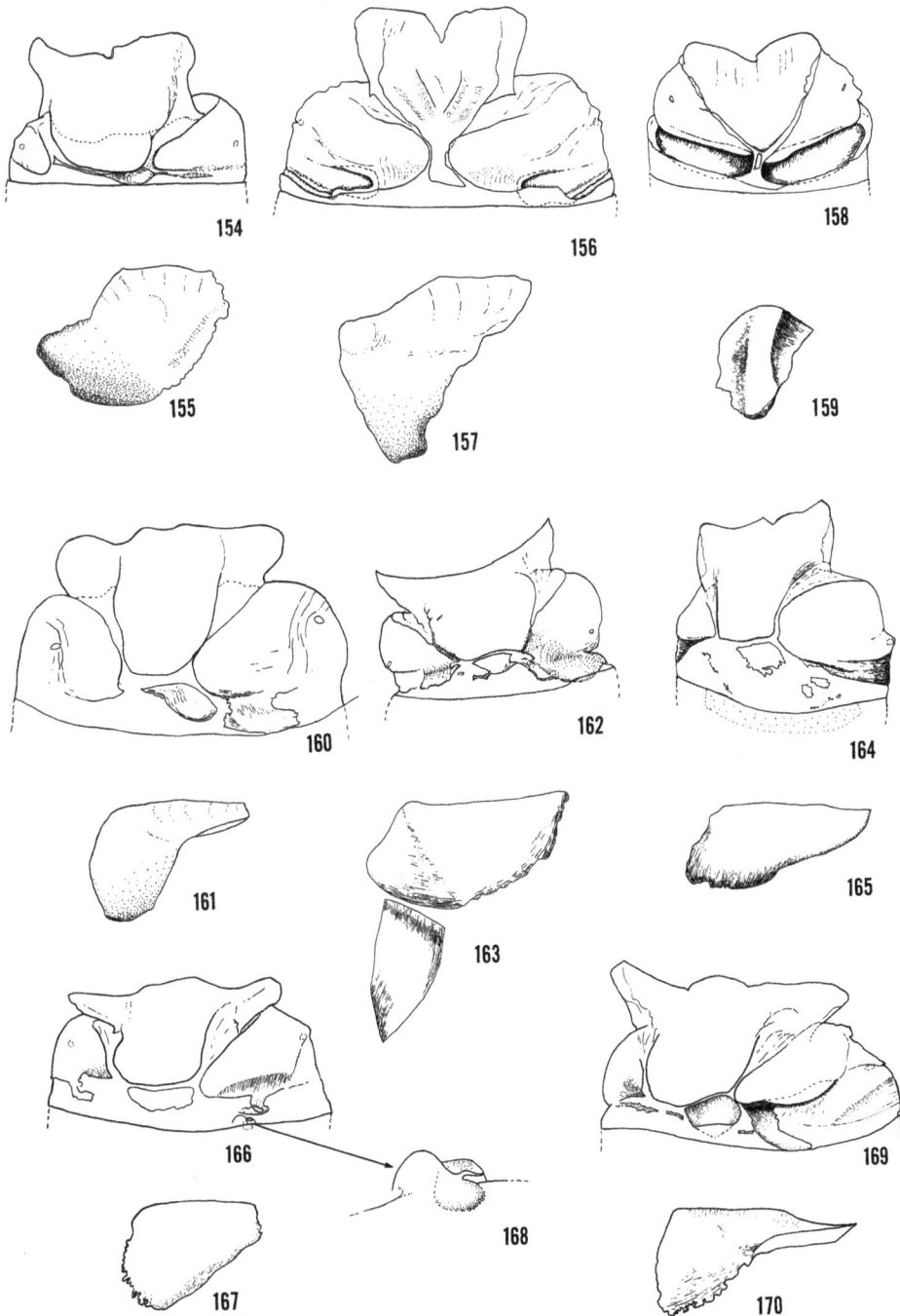

Figures **154-170**. Abdominal sternite VII and sclerite of lamella postvaginalis in females. Figs. **154-155.** *M. thyridia*, Río Morichal Largo, Venezuela; **156-157.** *M. durcata*, Rancho Grande, Venezuela; **158-159.** *M. aurifera*, Sarayacu, Ecuador; **160-161.** *M. bodoquero*, Pebas, Perú; **162-163.** *M. coerulescens*, Rancho Grande, Venezuela; **164-165.** *M. orichalcea*, Pará, Brazil; **166-167.** *M. tarsispecca*, Río Chipiriri, Bolivia; **168,** enlargement of intersegmental pocket; **169-170.** *M. immanis*, Chulumani, Bolivia.

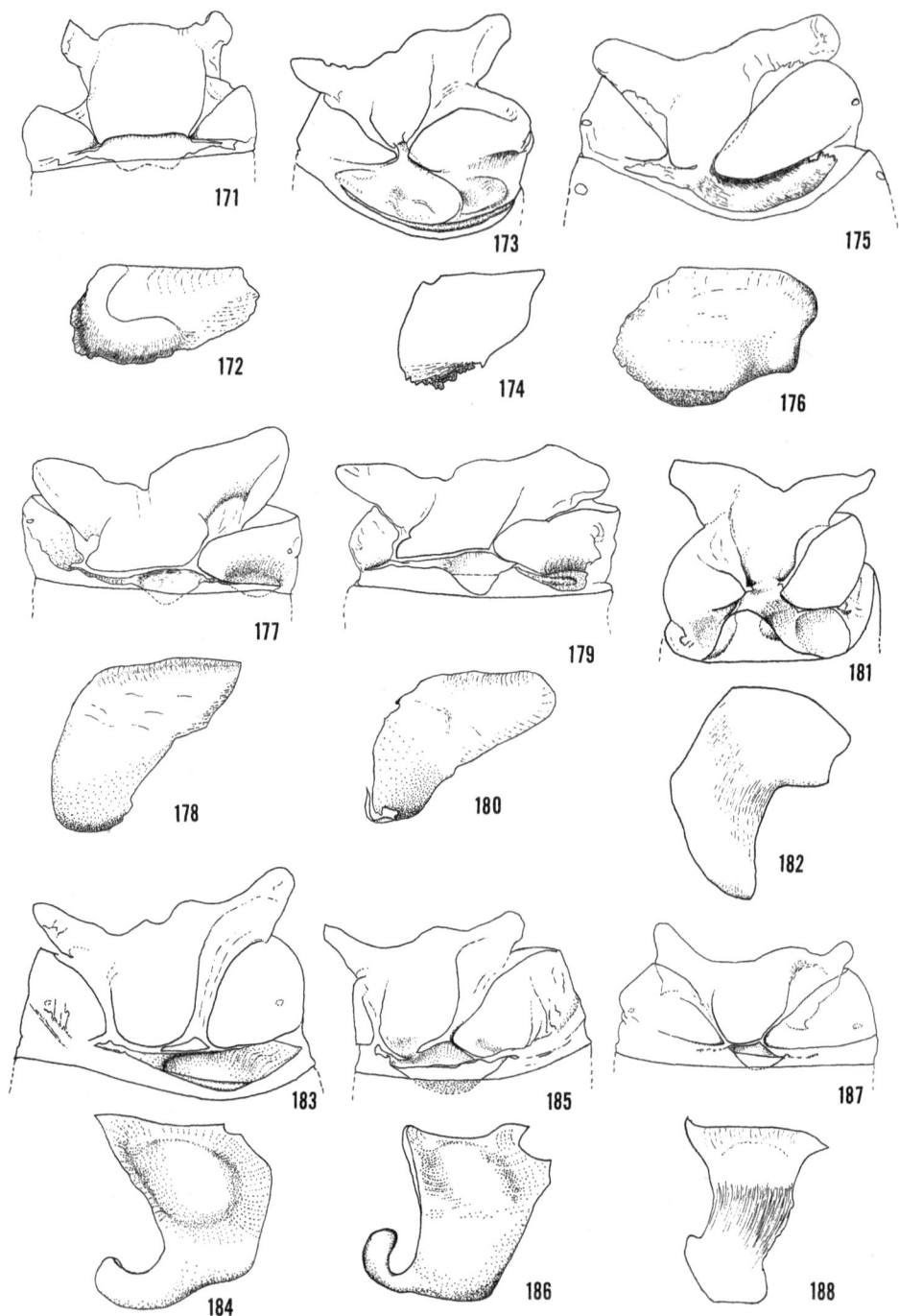

Figures 171-188. Abdominal sternite VII and sclerite of lamella postvaginalis in females. Figs. 171-172. *M. zongonata*, São Paulo de Olivença, Brazil; 173-174. *M. cyanea*, (Brazil?); 175-176. *M. cupreipennis*, holotype, Brazil; 177-178. *M. megacybe*, Castro, Brazil; 179-180. *M. ferrea*, holotype, Espirito Santo, Brazil; 181-182. *M. leucostigma*, Asunción, Paraguay; 183-184. *M. pelotas*, Pelotas, Brazil; 185-186. *M. mormo*, Rio Mucuri, Brazil; 187-188. *M. bestia*, Imbarie, Brazil.

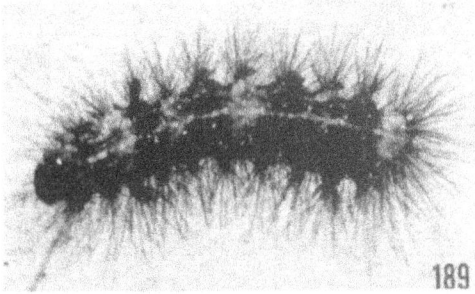

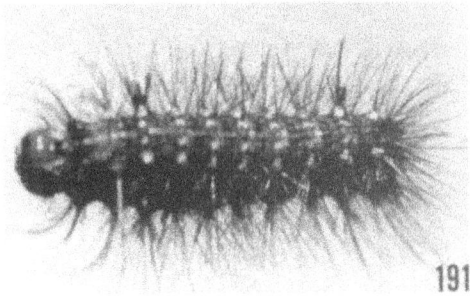
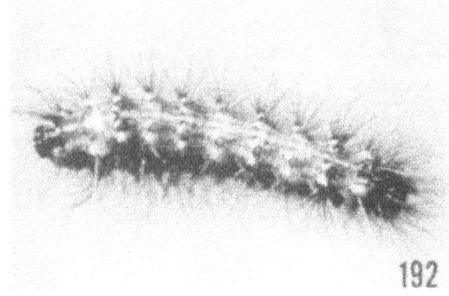

Figures **189-192**. Larvae, reared on artificial diet: **189.** *M. thyridia*, lot 12E75, Guatopo Nat. Park, Venezuela; **190.** *M. thyra*, lot 17E75, Guatopo Nat. Park; **191.** *M. coerulescens* (=*yepezi*), lot 15E75, Guatopo Nat. Park; **192.** *M. coerulescens*, lot 1F75, La Soledad, Venezuela; Figs. **193-194**. Pupa, *M. coerulescens* (=*yepezi*), actual size, 13 mm: **193**, lateral view; **194**, ventral view.

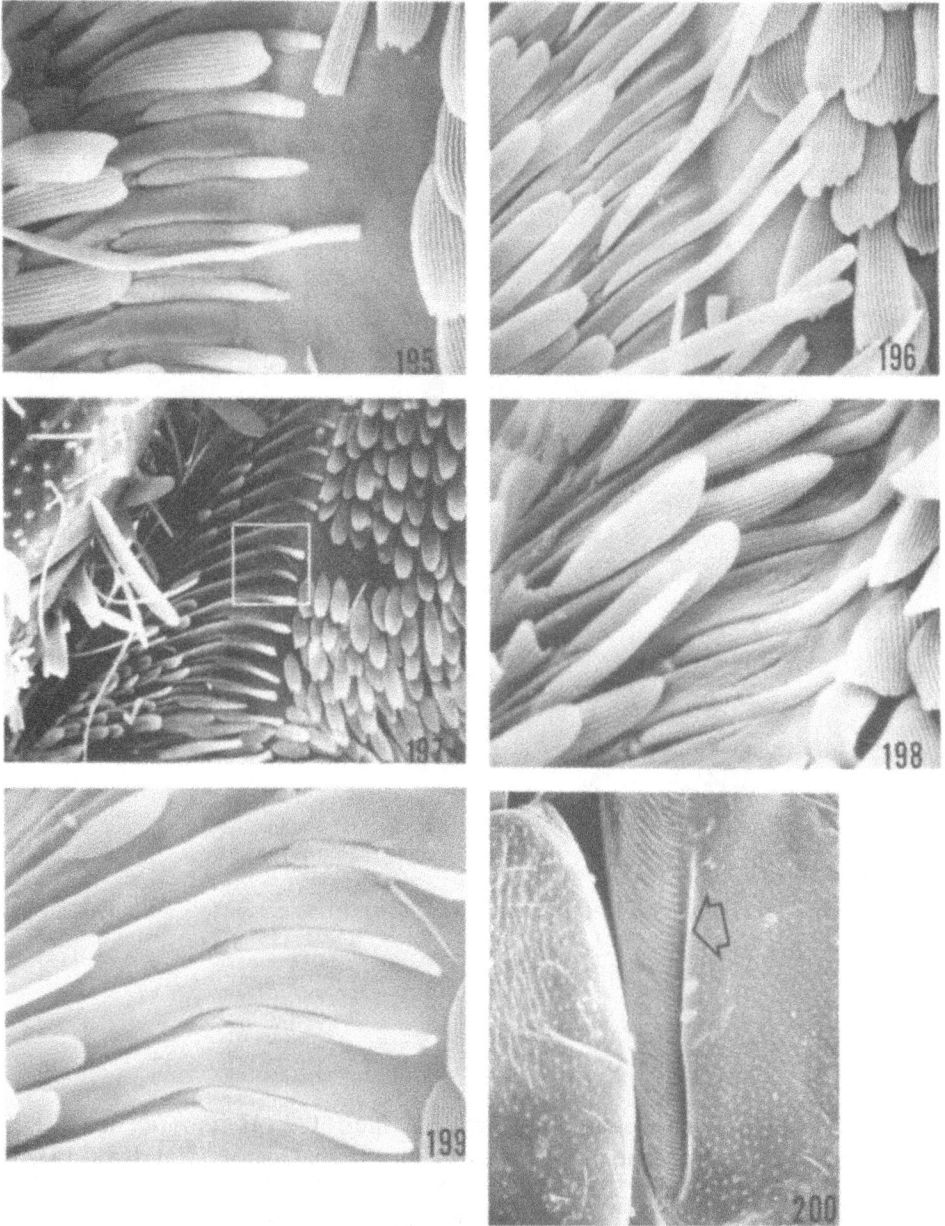

Figures **195-200**. Tymbal organ showing differences in size of microtymbal hairs: **195,** *M. chrysitis,* 580X; **196.** *M. lades,* 580X; **197.** *M. adonis,* 155X; **198.** *M. thyra,* 720X; **199.** *M. adonis,* close-up, 550X; **200.** *M. chrysitis* with scales removed; arrow = artifact from drying procedure.

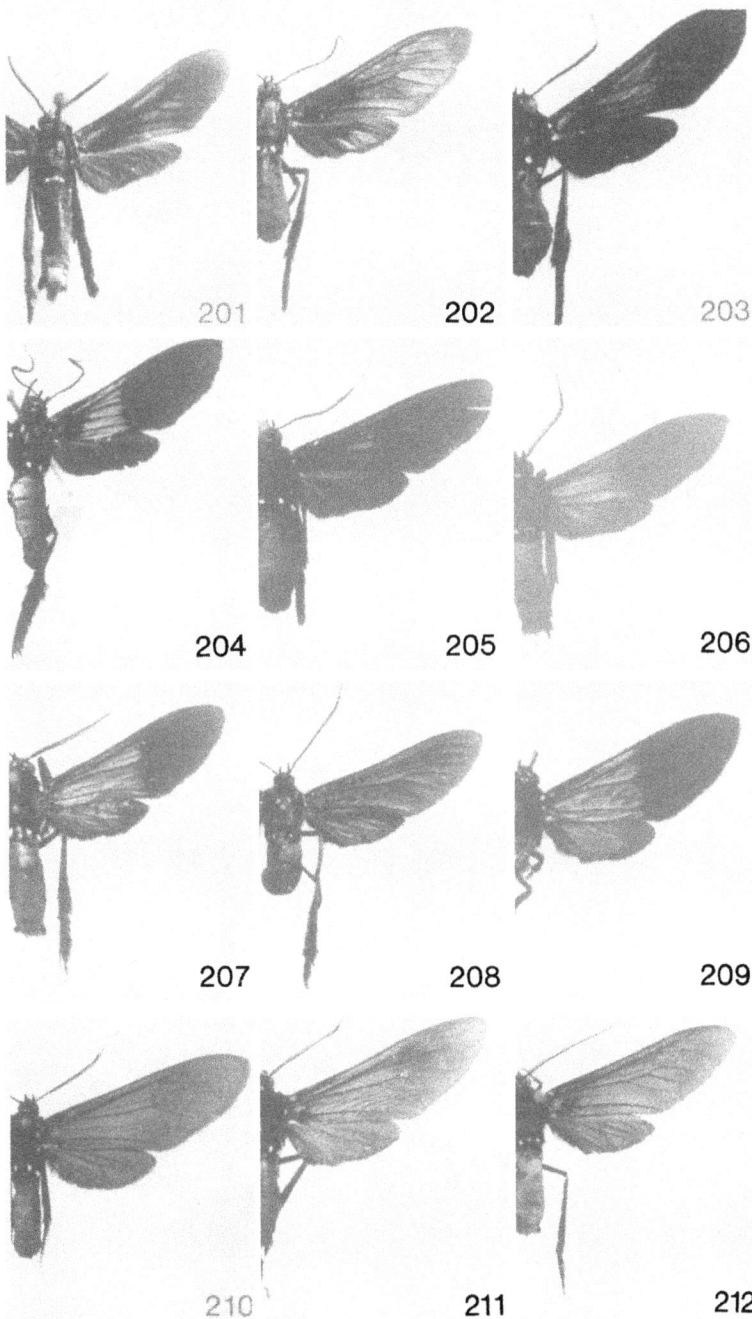

Figures **201-212.** Adults: **201.** *M. adonis* (=*cinyras* Schaus), male, Unión Juarez, Mexico; **202.** *M. adonis*, female, Rancho Grande, Venezuela; **203.** *M. chrysitis*, female, Jacala, Mexico; **204.** *M. iole*, female, San Vito, Costa Rica; **205.** *M. iole*, variant female, San Vito, Costa Rica; **206.** *M. semiviridis*, male, Sto. Domingo de los Colorados, Ecuador; **207.** *M. oponiensis*, holotype male, Alto Río Opón, Colombia; **208.** *M. coerulescens*, large blue, variant female, Pichis and Perene, Perú, (cf. Figs. 218 and 219); **209.** *M. cabimensis*, variant male, Nare River, Colombia; **210.** *M. immanis*, female, Yungas, Chulumani, Bolivia; **211.** *M. habroceladon*, holotype male, Cochabamba, Bolivia; **212.** *M. tarsispecca*, holotype male, Sarampiuni, San Carlos, Bolivia.

Figures **213-224.** Adults: **213.** *M. lades*, neotype male, Régina, French Guiana; **214.** *M. lades*, female, Régina, French Guiana; **215.** *M. thyra*, male, Wineperu, Guyana; **216.** *M. thyridia*, female, Caño Mariusa, Venezuela; **217.** *M. thyridia*, male, Tarapoto, Perú; **218.** *M. coerulescens*, male, Las Lajas, Tachira, Venezuela; **219.** *M. coerulescens* (=*yepezi* Förster), male, Rancho Grande, Venezuela; **220.** *M. aurifera*, male, Rancho Grande, Venezuela; **221.** *M. ancaverdia*, holotype male, Lima Perú; **222.** *M. durcata*, holotype male, Aroa, Venezuela; **223.** *M. bodoquero*, holotype male, Morelia, Río Bodoquero, Colombia; **224.** *M. imbellis*, male holotype, Iquitos, Perú.

225 226 227

228 229 230

231 232 233

 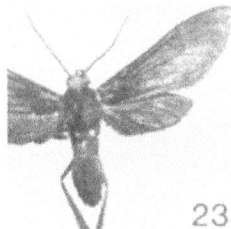 

234 235 236

Figures **225-235.** Adults: **225.** *M. pelotas*, male holotype, Pelotas, Brazil; **226.** *M. orichalcea*, male holotype, Pará, Brazil; **227.** *M. zongonata*, male holotype, São Paulo de Olivença, Brazil; **228.** *M. bestia*, male holotype, Imbarie, Brazil; **229.** *M. cupreipennis*, male, Hansa Humboldt, Brazil; **230.** *M. cyanea*, male, Hansa Humboldt, Brazil; **231.** *M. mormo*, male holotype, Rio Mucuri, Bahía, Brazil; **232.** *M. leucostigma*, female, Pelotas, Brazil; **233.** *M. megacybe*, male holotype, Alto da Serra, Santos, Brazil; **234.** *M. cabimensis*, male, Cerro Campana, Panamá; **235.** *M. melanopeza*, male holotype, Pan de Azucar, Pasco, Perú; **236.** *M. ferrea*, male, Santa Teresa, Espírito Santo, Brazil.

www.ingramcontent.com/pod-product-compliance
Lightning Source LLC
Chambersburg PA
CBHW080240270326
41926CB00020B/4319